W9-BAU-635

# Mice in the Freezer, Owls on the Porch

# Mice in the Freezer

*The Lives of Naturalists*
*Frederick & Frances Hamerstrom*

# Owls on the Porch

## Helen McGavran Corneli

*Foreword by* George Archibald

The University of Wisconsin Press

The University of Wisconsin Press
1930 Monroe Street
Madison, Wisconsin 53711

www.wisc.edu/wisconsinpress/

3 Henrietta Street
London WC2E 8LU, England

Copyright © 2002
The Board of Regents of the University of Wisconsin System
All rights reserved

5     4     3     2     1

Printed in the United States of America

Library of Congress Cataloging-in-Publication Data
Corneli, Helen McGavran
Mice in the freezer, owls on the porch: the lives of naturalists
Frederick and Frances Hamerstrom / Helen McGavran Corneli.
p. cm.
Includes bibliographical references.
ISBN 0-299-18090-5 (cloth: alk. paper)—
ISBN 0-299–18094–8 (pbk.: alk. paper)
1. Hamerstrom, Frederick. 2. Hamerstrom, Frances.
3. Naturalists—United States—Biography.
4. Conservationists—United States—Biography. I. Title.
QH26 .C63 2002
578′.092′273—dc21

2002002338

For Kip
Who took me to the sand counties
and turned my life around

The whole human race spends too much emotion on itself. The happiest and freest man is the scientist investigating nature or the artist admiring it, the person who is interested in things that are not human. . . . We can best fulfill . . . humanity's claims by keeping our emotional sanity, and this by seeing around and beyond the human race.

—*Robinson Jeffers*

# Contents

# Foreword

In this compelling book, Helen Corneli salutes two of America's most fascinating and accomplished field naturalists, Frances and Frederick Hamerstrom. It's a heartwarming account, full of unpublished stories and insights about two people who are well known in both scientific and popular literature of the last half-century. The author, a gifted writer who knew them well, taught English at the University of Wisconsin–Stevens Point. She provides her readers with an eloquent account of the innovative and sometimes unorthodox lives of one of the twentieth century's most remarkable couples.

As a graduate student at Cornell in the late 1960s, I met raptor biologists who gathered at the Laboratory of Ornithology to study birds of prey under the direction of Dr. Tom Cade (founder of the Peregrine Fund). They told me about these pioneering ornithologists from the sand country of central Wisconsin. Fellow graduate student Jim Grier, later one of the nation's leading authorities on eagles, introduced me to the Hamerstroms when they presented an impressive paper at Cornell about the effects of DDT on harriers and kestrels. Several decades of data comparing the ability of these DDT-sensitive raptors to breed before and after being exposed to insecticides provided some of the ammunition that led to the banning of DDT in the United States. Fran Hamerstrom also brought to Cornell her two golden eagles, conditioned by her to trust her, and presented them to Jim on captive breeding loan. Through the Grier-Hamerstrom efforts, in 1972, the first artificially inseminated golden eagles in the world were produced.

As I was completing my stay at Cornell in 1971, Ron Sauey from Baraboo, Wisconsin, was beginning his graduate work at Cornell. Ron felt that his experiences with the Hamerstroms—as a volunteer research assistant one summer during his high school days—changed his life. Ron had been raised in a loving and economically privileged family. His parents were aware of his interest in nature and, although they hoped he might enter the world of commerce, they were delighted that the Hamerstroms might provide something his family could not. Living in the Hamerstrom home with its outdoor toilet, hand-pumped water, and lack of central heating, helped Ron see the world from a completely different perspective. The set of values articulated by Aldo Leopold and lived by his students, the Hamerstroms, became his inspiration to pursue a career that involved the study and conservation of birds. Soon, in 1973, Ron and I cofounded the International Crane Foundation in central Wisconsin. Ron was but one of several hundred young people whose lives were changed by their experiences with Fran and Hammy who opened their minds, their fieldwork and their home to strangers. (A potato and a plucked pigeon in a paper bag was one of their favorite Christmas gifts to friends.)

It was always inspiring and entertaining to visit the Hamerstroms. You never knew what might happen in a home with a full-winged great horned owl and a refrigerator containing fresh and frozen roadkills. One friend recalls having the owl land on her head while seated at the table. Fran commanded, "Joyce, stand up very slowly. Now, sit down very quickly." As the roosting spot dropped, the owl flew to another. On another occasion, Fran placed the jacket of her latest popular book on a volume of biochemistry. She passed the book to a visiting professor and asked how he liked her new work. It took him several minutes to realize the practical joke. Fran and Hammy waited in eager anticipation for the revelation. They loved good fun.

It was a thrill for me to be able to introduce Russian raptor biologists and my close friends, Vladimir Flint and Alexander Sorokin, to the Hamerstroms. Both parties had written extensively about falconry and they bonded instantly. One evening at my home, Fran and the Russians viewed a videotape of Fran's appearances on *Late Night with David Letterman*. Fran had not previously seen her performance (there was no TV in the Hamerstrom home), so both she and the Russians shook with laughter every time Fran pointed and announced, "Look what she's doing now!" She enjoyed being, and even more watching, Fran Hamerstrom.

One memorable spring evening, Hammy, Fran, and I sat beside their potbellied stove, drank a little red wine, and chatted into the wee hours of the night. Then Fran led me to my unheated upstairs bedroom. The window was open and the moonlight streamed into the room. In the nearby forest a ruffed grouse drummed. Lying there, I had a feeling of ecstasy so profound that I fought sleep. The glow from the company of the Hamerstroms, the moonlight, and the grouse all met in a dreamlike state that will remain one of my most cherished memories.

But the two of them had a special ability to relate to many audiences, including a small group of their well-heeled friends and other influential people in Milwaukee who founded, with Hamerstrom recommendations, the Society of Tympanuchus Cupido Pinnatus, Ltd. The annual meeting of the society was a grandiose cocktail party in Milwaukee just before Christmas. The prairie chickens boomed and danced on a screen while the Hamerstroms circulated among the millionaires. The business meeting occurred during a lull in the roar when Tympanuchus, the Society's president, called the meeting to order, the treasurer announced the amount of money that had been raised to buy land for the chickens, and Tympanuchus swiftly called for adjournment. Thus the society saved the critical nesting habitat for a remnant population of about 1,000 prairie chickens.

Although well known in the ornithological circles for their classic studies of the prairie chickens, harriers, and kestrels, they are best known through a series of widely popular books penned by Fran. (*Walk When the Moon Is Full* is one of the most cherished books for children in Japan.) And in 1982, when I married a lovely Japanese lady, Kyoko, the Hamerstroms were pleased because they felt interracial marriages were one of the best ways to improve international relations. Fran and Kyoko enjoyed one another, and Fran sometimes stayed at our home after Hammy passed away. One evening before retiring she asked Kyoko for a small empty coffee can. Kyoko was mystified. "Kyoko dear, it's much more convenient than walking all the way to the bathroom." Fran was practical.

Fran and Hammy were so close that we all feared she might wither after he was gone. But when she returned to Wisconsin, she seemed invigorated. She told me that death was a part of living and that the living must continue to live. Fran had always wanted to study in the bush firsthand the hunting practices of the Pygmies in Africa and the Indians in South America. Several winters during her eighties, she lived with traditional hunters in the jungles. She communicated by sign, lived in a little tent in their villages, and followed the hunters as they silently roamed the forests in quest of prey. She remained an adventurer even in the last three years of her life when afflicted by lung cancer. Nonetheless, her winters continued in the jungles of Peru and summers at the homestead in central Wisconsin.

During her last summer, we invited all the Hamerstroms' former students, fondly known as "gabboons," to visit Fran at the guesthouse at the International Crane Foundation. About thirty came from faraway places across the continent to spend quality time with Fran. She delighted in zooming around ICF in a golf cart and introducing her friends to the cranes and the prairies. On the final night we shared a picnic around a fire at Aldo Leopold's shack just a few miles down Shady Lane Road.

Hammy and Fran had spent most of their lives in the sand counties living in harmony with nature and teaching others as Aldo Leopold had taught them. They challenge us to do the same.

George Archibald
International Crane Foundation

# Acknowledgments

There would have been no such book as this without Fran, who approached me about writing it a few months after Hammy's death. (I still wonder whether that modest, reticent man would have approved of the project.) Without the generous cooperation of the Hamerstrom and Flint families, I could have accomplished little. I have tested the excellent memories of Dorothy Ann and Putnam Flint; they could not have been more helpful. And were I able to bestow sainthood, Elva Hamerstrom Paulson would be a prime candidate. She was always encouraging, and her artist's eye, patience, and industry were invaluable.

Beyond that essential circle, I am especially grateful to those who provided positive suggestions. Ruth Hine, from the very first, pointed me toward the essential themes and later helped me trim and order a great shaggy mass of information. James B. Hale has been unfailingly responsive to questions on many levels, as well as sharing with me many lively glimpses of my multi-faceted subjects, complete with pungent phrases. Paul Hass gave me the most delicate and creative introduction to professional editing that I could ever have imagined. He honors the profession and I was fortunate to have had some months of answering his charming, tactful, and incisive letters.

The successive series of sensitive manuscript readers slapped down, kneaded, shaped, and proofed my textual sponge. That process turned the rising mass into what I hoped it would be—a fragrant, crusty loaf of a book. Special thanks to Steve Holmes, who with care, candor, imagination, and diligence, helped me connect the bones, as it were, of the subject. May all writers find such a reader.

And now rise up a crowd of Hamerstrom colleagues, friends, neighbors, acquaintances, members of bird-related organizations, and gabboons. I already knew some of them and got to know many more. It didn't surprise me that they seemed delighted to talk of their times with Fran and Hammy. It was a cross-section of humanity I met, generous with their memories, frank in their pronouncements, and appreciative of my desire to re-create the contradictions and complexities of the Hamerstrom years. I thank each one. Particularly important to my understanding of Hammy and Fran were Os and Mary Mattson, Ray Anderson, Joe and Sheila Schmutz, Dan Berger, and Jim Weaver.

The splendid group of talented and diligent professionals at the University of Wisconsin Press gently approved the new approaches, experiments, and demands that preparing a credible and attractive book requires. Steve Salemson's patience and openness were particularly helpful.

My husband, Kip, and my children, each one of whom knew the Hamerstroms, have all played important parts in these years of growing a biography.

# Mice in the Freezer, Owls on the Porch

# 1

# Prologue

*And although nothing much can be seen through the mist, there is somehow the blissful feeling that one is looking in the right direction.*
—Vladimir Nabokov, *Speak Memory*

Well before dawn on a marrow-chilling morning in April 1961 I found myself with a reporter friend, kneeling in a four-by-four-by-six-foot canvas box in a central Wisconsin marsh and peering out of the vision slit on one side. I was shivering and ruefully remembered my husband's stern words of the evening before.

"Wear your long johns and ski pants; and at least two pairs of socks. Put a sweater over that wool shirt. Slip a plastic bag between your socks and boots—it may help. Take these extra mittens," he warned. "These Wisconsin spring days are *cold!*"

"I'll be fine," I assured him airily. Now I was freezing. The bedroom in the Hamerstrom house where I had slept was unheated, the rattletrap van that a silent youth maneuvered over a rutted mud road to the marsh had no heater, and the walk to the blind from the roadside where he dumped me was icy. My fingers were numb, my feet ached with the cold. We had been cautioned to be still; I dared not stamp my feet to warm them. The space was constricting, the chill penetrating.

Barbara, a stringer for the county paper, had asked me to "go booming" with her. I had met the Hamerstroms—cordial, interesting people, I thought, though perhaps a bit eccentric, so I readily agreed. Now, not fully awake, I tried to remember what Dr. Hamerstrom had told us the previous night. There would be cocks, with something called "bellows" and "long flange feathers" on the neck, and hens; we

3

were to watch what happened in any encounters. There might be matings. The cocks would display—hopping, jumping, and even fighting. On site, it seemed an unlikely proposition. "How will we ever see a bird? It's still dark." I whispered to Barbara.

"Shhh!" she breathed, "they should come soon. You have the paper for the map, don't you?" I fumbled in my jacket pocket. Suddenly, startlingly, there was a reverberant, unearthly sound echoing, apparently right above us—around us—or was it beside us? Barbara pointed and I swung the unfocused binoculars. Where was it? When I put the binoculars down I found in the thinning dark a creature making odd, hopping jumps with its head held low. In the dimness I wondered: It couldn't be a rabbit—but there were odd protuberances at the head. Then I recognized the inflated orange sacs. It was a prairie chicken cock.

"There's another—a hen—I think," Barbara murmured.

"Can you see a band? Here are the binoculars."

"Wait . . . yes! I think there are bands on both legs!" Then, after a long moment, "But I can't make out the numbers."

The sound was unforgettable. Like whale song, like the cry of cranes as they float to their evening roosts, like the chanting of Tibetan monks, it held overtones of a strange and distant past. My ears might well have been in my chest. The echoing resonance did not continue long: the single cock, with no rival to stimulate display, soon quieted. With only four birds in all, the group began to resemble domestic chickens, casually strolling, scratching, or pecking at the ground. One by one, they wandered away. We waited, sketched a simple map, chatted quietly, and at full light, tipped the blind and walked to the road where we waited for the van. I felt strangely let down. The exercise we had endured seemed almost pointless, and our debriefing back at the house with Dr. Hamerstrom was short. He sat at the table, pencil in hand and record sheets before him. He glanced at the map. "Thank you," he observed. "We haven't had much activity at that blind this year. There's been a lot of clearing and plowing nearby." Then he turned to the next observer. I didn't know it then, but we had been given the customary easy assignment, usually passed along to newcomers. Although the morning had been interesting and Barbara got her story, I had no thought that "going booming" would be the start of not only a significant friendship but also of a compelling education.

My husband, Kip, our children, and I had moved to eighty run-down acres near Hancock, Wisconsin, in 1959. A graduate student in the department of horticulture at the University of Wisconsin, Kip had responded to the lure of farming on the developing "golden sands." We lived about five miles south of the Hamerstroms in an old farmhouse whose uninsulated walls filled with condensation ice in the winter. On sunny spring days I used to go outside to warm up.

The culture shock was worse than the cold. We had come from three contented years in cosmopolitan Madison to what appeared to be a northern Appalachia. Attitudes here were different. Scorn greeted the wife of the owner of the Hancock lumberyard, who both clerked and kept the accounts. But, "She's got a housekeeper!" said a neighbor. "She's too fine for housework!" A housewife might earn the accolade: "You could eat off of her floors!" Thrift was both necessary and valued. In 1962, when a stray cow, mangled by a car on Highway 51 near the university experiment farm, was dragged away and buried, the director's wife repeatedly heard the baffled query: "Why didn't you butcher that cow?" Hundreds of jars of home-canned food filled cellars, children labored for hours in the sweltering summer fields, picking green beans and cucumbers for cash to buy school clothes. School was regularly canceled for potato harvest and hunting season.

Gender roles were fixed: men did the field work; women gardened, cooked enormous meals, heated water and washed dishes in a pan three times a day, canned, cleaned, scrubbed laundry, sometimes with washboard and tub, and then helped with the milking. A sensible woman offered her opinion only in the privacy of her home, her Homemakers club, or, perhaps, her church circle. Husbands too obviously influenced by their wives were regarded with disdain. The telephone party line circulated local news, but a purportedly more reliable source of information was found at the local gas stations in Hancock or Plainfield where a group of men gathered nightly to swap stories while "the wife" did home chores. At social gatherings, men retreated to the porch or basement to discuss machinery, livestock, and the eternal complaint, "Ain't it dry!" Unusual people and new ideas were greeted with guffaws or suspicion, and newcomers were automatic outsiders.

Our close neighbors, however, were helpful and friendly. Early on, one suggested that we should meet the "bird people" who lived close by. We called; we were invited in and offered coffee or wine. It was good to see walls of books and piles of *New Yorker* magazines on the floor. Wildlife art on the cracked plaster walls and a motley collection of gear betrayed an interest in birds, but the conversation was not about prairie chickens. We didn't even know, at the time, that the species was threatened. They—Frederick and Frances Hamerstrom— welcomed us at our level, asking kindly and neutrally about our intentions and the potential of farming in the area.

We learned about our new friends because of Fran's expertise with birds. When our children found an abandoned hatchling field sparrow, we called her. "We can't get it to eat."

"Mash hard-boiled egg up with a bit of milk and soft bread."

Meanwhile, we discovered that the promised gold of the sands would come only with considerable emendation of the soil. My husband taught in the consolidated school district, and presently, responding to the quality of the high school English program, I visited the English department at the college at Stevens Point to find out about becoming certified to teach at the high school. The chairman suggested I teach at the college instead, and I did. We built a better house, proceeded to dig a pond in a low corner, and settled in.

When someone gave us ducklings for Easter, we found that they would drown if soaked, as one batch did in the basin the children had thoughtfully supplied for the balls of fluff. We took replacements to our pond to swim, thinking we could lure them out before dark with feed. A sudden thunderstorm blew in. They ignored the feed we offered. "We'll have to catch them," said Kip. Ducklings are instant experts at diving: half an hour of furious cartoon-like pursuit left us, as dusk fell, frustrated, breathless, and empty handed. I called Fran. She drove straight down, a long flashlight in hand.

"You've got a butterfly net?" She eyed it appraisingly. "That'll do. Come on." She kicked off her shoes and moved quietly into the shallows. Click! As a duckling froze in the strong beam of light, she moved quickly through the dark and scooped it up. Two more clicks, two more quiet moves, and she got another. "Now, you try it!" Before long, we had them all.

She nurtured our friendship. In December she invited us up to see the flickering candles on their old-fashioned Christmas tree. In the cool, shadowed room she lit each one with a match, and, after a few minutes, she let each child pinch two candles out.

All the while, we were learning about Fran (pronounced Frahn, with the long Boston *ah* sound, as she insisted it must be), and Hammy. We found that they had been students of Aldo Leopold, whose *Sand County Almanac* I knew and loved. We saw them living the credo he proposed. We became aware of their efforts to save the prairie chicken in Wisconsin from almost certain extinction. Occasional reports of their demonstrations of astute interaction with policy makers and the public appeared in local papers.

We began to recognize the significance of the prairie chicken, one of several kinds of grouse. It is a plump, deep-chested, swift-winged bird that thrived in the days before settlers plowed up the prairies of southern Wisconsin. The shrinkage of its original wide range made this wildest of wild creatures, as hunters dubbed it, a major actor in a twentieth-century conservation drama. We were an audience, in the peanut gallery to be sure, of the drama of preventing this bird's extinction in Wisconsin, and the Hamerstroms were the stars.

We knew them primarily as compatible friends. I've never really known why they were so good to us. Perhaps it was in recognition of our loneliness and need for intellectual companionship, or perhaps they genuinely enjoyed and cared for us. One winter evening Fran called, a bit urgently. "Helen! Are your children all right?" I was perplexed. I looked around. "I think so, Fran. They're doing homework at the dining room table." She explained. She had heard of two children who had drowned, ice-skating on a farm pond. She had feared they were ours. "I just love your children," she said.

It might have been the pond that cemented our friendship. Ours, dug at a low corner where water stood in spring, was our celebration when, after a disastrous first year, we finally made a little money. We swam in it in the summer, the children taped spring frog choruses there, and it was splendid for ice-skating on winter days. In 1962 Hammy called and asked about it. Kip told him of the assistance for building ponds available through the Agricultural Stabilization Office at the county seat. "Thank you, Kip," he said, "I'll look into it." The next call was an invitation. "Come up and see our pond."

It was in their woods, lovely, still, and deep. Only in the 1990s did Fran reveal that after driving by our pond she went home to announce, "The Cornelis just made a pond. I want one." That had prompted Hammy's call to us, but he decided, "None of those government subsidies for me." He would not budge. Not long after, old Mr. Knaack, who lived in the marshy country nearby, drove into their driveway in his rattletrap Chevy pickup.

"Say," he asked Hammy, "Can I buy some of them young birches from you?"

"What do you want them for?"

"People want birch trees. I dig 'em and plant 'em. Most of 'em live." Fran knew of Hammy's plan to burn those birches. She held her breath. "I don't really want to sell . . . " he said slowly, "But you can talk to my wife."

She sold the birches for $250. Then she went back to Hammy. "Look at the money I just made for the pond." Exasperated, he sighed, "I wash my hands of the whole thing." He did not know of the additional $250 she had painstakingly saved. She applied for and got the subsidy. The subsidy, her savings, and the birch tree money gave them their $1,000 pond. Years later, as they finished their afternoon dip in the cold, clear water, Hammy turned to her. "This pond," he said, "was the best investment we ever made."

Our friendship with the Hamerstroms grew closer as their children, Alan and Elva, grew up and moved away. There were dinners at their house and ours, and we four became fast friends. Fran might ask us to house a visiting relative or ornithologist; very occasionally, she asked to take a hot bath in our modern bathroom or to bring a guest down for a sauna.[1] And then, after Kip became an enthusiastic small-plane pilot in the seventies, she called us.

"Kip, do you need an excuse to fly your plane somewhere?"

"Anytime!" he responded. "But the weather has to cooperate. I'm VFR—that's visual flight only, you know."

Fran explained. She wanted to observe growing owlets, but to raise an owlet requires a permit. She got one and arranged to receive three hatchlings from Patuxent Research Station in Maryland. A few hours in a commercial flight would bring them speedily to Minneapolis, but the four-hour drive back to Plainfield might be fatal to the fragile chicks.

In due course, the four of us flew to Minneapolis. Hours before the owls' arrival at the airport, we landed at the light plane terminal. Kip and I found a Chinese restaurant and ordered tofu and vegetables. The carnivorous Hamerstroms tasted it. "Awful!" said Hammy. "How can you eat this stuff?" Shortly, at the courier's gate, they took the insulated, ventilated wooden box he handed them and rushed back to our terminal with one-, two-, and six-day-old owlets, alive and hungry.

An owl provides her young with regurgitated flesh: Fran did the next best thing: she took a plastic bag from her purse and placed several strips of raw meat from it in her mouth to warm and moisten. Hammy held the owlet box, from which she picked up a hatchling.

A young owl is the ugliest creature you can imagine. A grotesque bumpy skull with an outsized beak dominates a scrawny body, sketchily dressed in down. Featherless and with closed eyes they are, quite simply, hideous. Fran took a strip from her mouth and touched it gently to the side of the owlet's head. It gaped. She slipped the meat in the beak, waited, and repeated the process. A young man in a blue pinstriped suit came to a sudden stop in front of her and stayed there staring, absorbed and mystified.

"What is that?" he asked. There was no answer. He asked more urgently. "What *is* that?" Fran glanced up; he moved a step closer. "What is that you are feeding?"

"A bahnowl," she said shortly, and put another strip of meat in her mouth. "What?" frowned the puzzled man.

"A bahnowl, a bahnowl!" she replied impatiently and returned to the owlet.

Puzzled, he turned to us. "What in the world is a bahnowl?"

"It's a barn owl," I said, emphasizing the *r*. "She's from Boston," I explained.

Hammy intervened. "This is a newly hatched owlet. We have to keep it fed and warm. We hope to raise it to maturity, but in these first hours, frequent feeding is necessary."

The enlightened man smiled and hurried on. Within minutes, we were in the plane and headed for home. The Hamerstroms sat in the back seat with the owls. I glanced back after takeoff: Hammy was holding Fran's hand, their smiling faces glowing. I wished then that others could know more of their spirit and their way of life.

Perhaps that experience was the seed of this book. Indeed, shortly after that I began to think of writing a profile of them for the *New Yorker*. After any particularly interesting experience up at the Hamerstrom house, I jotted notes, which I carefully stored in a file in my desk. That was as far as that project got, but the summer after Hammy's death in 1990, I happened to mention to Elva my regret that I had never pursued my intention.

The next day I heard Fran's throaty Boston accent on the telephone. "Helen! I hope you'll think about writing Hammy's biography. It must be done." I protested that I knew far too little. "I can tell you lots of stories about Hammy," she replied. "And there are drawers of material right here."

I was curious about Hammy. What had formed this man who combined courtliness with integrity, discipline with grace, principle with practicality? How did he become so unabrasively honest? I knew then that their story was a kind of case history of a conservation conflict, one that might illustrate common principles. It was enough to get me started.

Fran took me on. She loaned me her overflowing folder of clippings, warning me that it must stay in the order I found it—a directive that she amended after I asked if she didn't simply want it to be chronological. Then she opened Hammy's office to me. Three almost impossibly tightly packed file drawers awaited inspection. Once I left one drawer slightly ajar. She took me to the offending file the next day. "Some mammal," she said calmly, "left that bottom file drawer open. Open drawers invite mice; mice destroy paper." And that—in her phrase—was that. She gave me practical guidance, listing a group of old friends and advising, "Call them as soon as you can," she said, "They may not be here for long."[2]

In those years immediately after Hammy's death, I spent hours hearing her stories and recording her answers to the questions I raised. She was patient and forthcoming, unafraid to say, "I really don't remember." She regaled me with reproductions of their conversations: "Hammy said . . ." or "I told him . . . ," and I wrote down these sometimes salty tidbits and then checked her facts. I needed a chronological narrative—the last thing on her mind. Even more, I wanted validation of both Fran's stories and those of other people, both of whose accounts might be mistaken, exaggerated, or even fictitious.[3]

Recording such material requires practical conventions. Where I heard an actual Hamerstrom statement, I simply quoted it. Should Hamerstrom-talk come to me from others, I record the name of the person who heard or wrote about it. If I quote from one of her books I give the source, but because she often told me slightly different versions of tales appearing in her books, I occasionally use the story in the form her son used to call "her story-telling mode."

I saw at once that Fran believed I could weave her stories about Hammy into a life story. I could not: he deserved more. Yet it took several years of effort to realize that any real biography of him must include her. Only after her death did I feel I could complete the work. This book is, however, the testament of just one of their friends—and they were friends and mentors to many. Any one of their colleagues, acquaintances, visitors, and conservationists of other kinds would have written a different account. As a neighbor and sometimes confidant, not as a biologist, I tell their story as one of the ordinary people they reached and informed. Because I see their speech and their writing as integral to their persons, I have not hesitated to use their own spoken and written words.[4]

Tributes to the Hamerstroms recognize his stature as a distinguished, quiet scientist, and honor her as a creative and flamboyant woman, a "great lady," well known in raptor research and as the author of popular books. I honor both of them for saving the prairie chicken and for exemplifying a way of life. Aldo Leopold's words come to mind: "Twenty centuries of 'progress' have brought the average citizen a vote, a national anthem, a Ford, a bank account, and a high opinion of himself, but not the capacity to live in high density without befouling and denuding his environment, nor a conviction that such capacity, rather than such density, is the true test of whether he is civilized. The practice of game management may be one of the means of developing a culture which will meet this test."[5]

Fran and Hammy, truly civilized citizens of today's world, leave a record of devotion, adventure, and achievement. The maturation of their romantic, creative, and scientific partnership deserves attention. Above all, this book is, in the perceptive words of another friend, "the story of relationships, between a man and a wife, between two biologists, and between researchers and the researched."[6] It is, essentially, a love story.

# 2

# The Complexities of Childhood

*Whoever inquires about our childhood wants to know something
about our soul . . . we love with horror and hate with an inexplicable
love whatever caused us our greatest pain and difficulty.*
—Erika Burkart, in Alice Miller's *For Your Own Good*

What inquirer can discover the hidden country of childhood? Hammy
was silent about his, but Fran mined her memories and arranged them
in *My Double Life,* a book that invites reflection.[1] The saying, "It's all
true, even if it didn't happen," applies, for her selective, dramatic account of the life of a poor little rich girl and its themes of rejection and
rebellion became her truth. Her nonconformist life, achievements, and
strongly held views so distant from those of her family-of-origin surely
rise from these early years.

Fran was the first of four children and the only daughter of a
prominent Boston family with old money.[2] Her father, Laurence B.
Flint (born in 1874) and his older brother, Carleton, were deserted
by their father when Laurence was just eighteen months old.[3] His
mother, of necessity, became an executive assistant, valued by successful lawyers for copying documents in her beautiful, clear handwriting, and for her diligence, accuracy, and discretion. The necessity of
earning a living caused Helen Flint to board her sons with a family
named Giles, and when Laurence was seven, she shipped him and his
brother to Germany and Herr Thümmer's boarding school near Dresden.[4] Her career flourished; she was able, in time, to build a "cottage"
in Hull, the four-story house overlooking Boston harbor that Fran

13

describes in her memoir.[5] The effect of this disruption of family life on the boys is unrecorded but could not have been all bad, since Laurence chose to make adult contact with Herr Thümmer in the years before World War I.

In due course the brothers came home. Laurence became an office boy for the Walter Baker Chocolate Company. His marriage to twenty-two-year-old Helen Chase in November 1906 freed him from the need to participate in the corporate world and to ignore the example of his father-in-law, William Leverett Chase. Chase was a merchant prince who owned the H&O Chase Bags and Bagging Company; he went to his office every day, and his predictable and ordered life epitomized success to his adoring daughter. Helen Flint had the last word on how her inheritance was to be spent, which may have complicated the customary adjustments required by newlyweds. The young couple's first child, Fran, arrived on 17 December 1907, and brother Bertram arrived on 3 July 1909.

Fran paints her mother as weak, her marriage a long series of concessions to an unreasonable husband. (Fran's own marriage gave her little experience of a difficult husband or of bearing two children in less than eighteen months under the eye of an apparently ruthless mother-in-law.) Mrs. Flint Sr. was demanding of her son and cruelly critical of her daughter-in-law—and, according to her grandson Putnam Flint—of the rest of mankind. Her son had to take her to the symphony every Saturday in the season: ("Helen will keep the children"). The dutiful Laurence, a bright, autocratic, and stubborn person, was a difficult husband. He was determined to pursue his own interests: travel and "criminology, photography . . . and making parts for his expensive automobiles," in Fran's accounts. In 1912, Laurence took his growing family and his 1906 runabout, "Merry Wheels," to Germany.

Putnam Flint provided me with a timeline on which to place Fran's general, selective vignettes. Between 1909 and 1912, Helen Flint spent two years at a Dr. Stedman's clinic. She suffered from severe postpartum depression, which may explain her mother-in-law's 1914 comment, when a recovered Helen became pregnant again with another son, Vasmer. Grandmother Flint wrote in her diary, "Isn't it a shame that Helen is pregnant again!"[6] Putnam, the last child, was born in 1919.

Laurence Flint, the "international criminologist," who was deemed by his daughter, Fran, "to be feared and avoided forevermore," 1930s.

As her diary records, in 1912 Grandmother Flint accompanied her son, at least part of the time, on an Edwardian grand tour. Trunks and servants were sent ahead, dressmakers frequented, residences established in several cities. Grandmother Flint's diary also indicates an unusually rigid set of expectations for her first granddaughter, another aspect of what would seem to be a poor environment for any development of a close mother–daughter bond between Fran and her own mother.

Meanwhile, Laurence developed a system of criminal identification for the German government, using ordinary forensics and the new science of fingerprinting. By Putnam's count, Fran's time in Germany

Helen Chase Flint and her daughter, summer 1908. The contrast between this formal photograph and that of Fran and her firstborn records social history (chapter 6). The difficulty of Mrs. Flint's situation at the time may be inferred by the evident distance between her and her baby.

before the family's return to Boston in 1914 could not have been more than two years—time enough to stimulate an innate talent for languages. She claimed that she spoke Bavarian, Saxon, High German, and French, and was "expected to speak them all correctly."[7] Whatever the case, her ensuing years with a German governess built on the language exposure she experienced in her early years.

After a year spent at Chesham, Granny Chase's elaborate house in Brookline, Massachusetts, the family settled in Milton in 1915. Fran seemed happiest "downstairs" with grooms, friendly cooks, and the maids and their suitors. She felt out of place in the "upstairs" world. She spoke with an accent; wore clothes that were different—smocked frocks from Liberty of London, "babyish" in her eyes—and found it hard to make friends. Most important, from a current point of view, she lived with adults whose child-rearing practices fit Alice Miller's dictum, "All advice that pertains to raising children betrays more or less clearly the numerous variously clothed needs of the *adult*."[8]

Strong-willed Laurence, conditioned by a Germanic upbringing and little experience of parental love, demanded respect and obedience from his children. True, he forbade the maids to spank their charges, but Fran repeatedly describes his boxing of ears (not hers). She was often sent to bed, shamed, or publicly reprimanded. Her parents assumed that the habits and skills they demanded of their beautiful daughter would grace the international hostess she was sure to become. Mr. Flint also expected his sons would be gentlemen sportsmen—though not one of them became an avid hunter.

Accustomed to deferring to his mother, Laurence Flint allowed her to select and hire a governess, Fräulein Taggesell, to educate her grandchildren. The elder Mrs. Flint seemed to always know best. She saw Fran as docile and "spiritual"; when Fran demonstrated a butterfly she was holding by the legs, her grandmother insisted that the insect was sitting willingly on her hand, for "little animals trust her." Fran tried to correct her and was sent from the room for her honesty.[9] A letter from Grandmother Flint, in a fine, legible hand, greeted Fran on her fourteenth birthday. It reads, "In a famous novel which you will someday read, a father says to his daughter: 'Never indulge in any pleasure which is not a pleasure to look back on.' It is a fine principle, isn't it?

You told me years ago that you tried so to live that people would love you. That also is very fine."

Although that report and a surviving conventional childhood letter by Fran do convey a wish to please adults and to "be good,"[10] her stories describe a need—and many opportunities—to find and exploit adult weaknesses. Her father demanded honesty from her, but he caused the servants to bend the law in minor ways. She had contempt for her grandmother's unrealistic view of her. At eleven or twelve, she was precocious in noting family tensions and firm in her judgments: "My mother is weak. She didn't stand up for me. . . . Fräulein is afraid. She is afraid of silly things like the dark. . . . My father is to be feared and avoided forevermore. I have tried prayer and found out it does not work. I am alone in this world."

Fräuta, "*my* governess," as Fran referred to her, was the exception. She "gentled" their family. Putnam, however, says stoutly, "She was governess to all of us." Fräuta's dictums, often startlingly universal and inflexible, sprinkled Fran's conversations well into her eightieth decade. ("Fräuta trained me. She said, 'Frances! If you have to revise your written work, you were not properly prepared!'" or "Fräuta taught me never to use soap on my face.") But Fräuta used soft answers, appeasement, and gentle interventions with her testy employer; Fran learned from her example and training, and long remembered her tact. In the early days of Hammy's ardent courtship, Fräuta discovered them "necking." To her charge, this sophisticated European only said, "Fran, you must be very careful to look after your health." Fran understood her gently stated concern.[11]

Fran at twelve was "distrustful, determined, and leading my first double life." That life included deception, hidden collections, a secret garden protected by barriers of poison ivy she planted, misbehaviors of minor and more serious kinds at school and dancing class, and above all, joy and fulfillment in nature study. She needed to experience the natural world, partly as a rebellion against and escape from the hypocritical adult world, but even more to delight in what she found. She records seeing her first edelweiss and golden eagles in the Alps; catching a hare in Dresden, petting the geese brought by steerage passengers onto the *Kaiserin Auguste Victoria,* or regularly sleeping outside after climbing out her window and training herself to waken early

A tranquil Fran in 1916 with Fräuta, her "beloved governess."

enough to get inside, put on her nightgown, and be in bed for Fräulein's wake-up embrace.

*My Double Life* re-creates her discoveries through a child's eyes: the harsh control of adults who commanded that she give back the goldfish a cook gave her, their refusal to listen, and their hypocrisy and perfidy.[12] Perhaps to compensate for failing to please men—first, her father and then adolescent males in the dreaded, hated dancing class— she determined to go her own way. She began smoking at age seven. She engaged in secret, persistent, even dangerous pursuit of answers to her strong curiosities. And when, unexpectedly, in late adolescence, she began to benefit from her beauty, a pattern of manipulation, daring, rebellious pranks, and secret independence was set.

A special account is that of 12 May 1916. Fran, not yet nine, was playing tennis in a children's tournament at West Chop, on Martha's Vineyard. It was a windy day. A truck roared up and stopped, and the driver asked for help fighting the fire on the nearby Great Plain. None of the vacationing city children volunteered. Fran's compressed account records a sense of shame, surely unwarranted: No adult wants

Sturdy eight-year-old Fran, barefoot and annoyed, in one of the hated frocks from Liberty of London.

an eight-year-old firefighter. Author Fran combined her childhood experience with a fact she must have learned later—that all the eggs for the year's crop of young were lost and a population of about 2,000 heath hens then declined to under 200, and the species went extinct by 1932. "I learned so young about the impending doom of a species, and later when I had grown up I helped to avert another doom."[13]

Her curiosity was active: she dissected a dead blue jay; she hurt her gums to get her way—earning visits to the Boston Museum of Natural History (her reward after visits to the dentist). She avoided dancing class by slashing her cheeks with brambles. But she acted to spare animal pain, healing a hurt falcon and mercifully killing a wounded kitten.[14] She had no idea that her fascination with nature would bring her a compatible husband and a rewarding career.

Fran would later regularly entrust her own children to her mother—the woman she painted as over-emotional and avoidant. Posey, as the children called their grandmother, joyfully cared for Alan and Elva for a number of summers. She bought their school clothes and took them to museums, ballet, and concerts. She constantly sewed and knitted for the family and corresponded with each one. She traveled from Milton to Chicago to pick them up in pre-school days, and regularly spent the hunting season in Plainfield to act as a stand-in parent. Elva now says, "She gave us a taste of fond, warm normalcy."[15]

Putnam, who had a mutually giving relationship with the father Fran depicted as a near ogre, maintains that Laurence Flint adored his daughter. To prepare her for the future awaiting a beauty, he once took her to a nightclub. He knew nothing of the surreptitious dates that had made Fran familiar to the club's serving staff. Luckily for her, she chose to deal with the awkward situation as a comedy, and her father assumed that the headwaiter recognized him when the man greeted her.[16]

Putnam saw his parents in a matter-of-fact way. "I witnessed on several occasions Posey laying down the law to Laurence. . . . Her [Fran's] attitude toward her parents has always troubled me. Where did she get her remarkable brain and style?" Putnam knew of Posey's generosity: she arranged for Fran to get $67 a month from a trust fund from her wedding day until May 1970—when the trust expired with Mrs. Flint. Fran, ten years older than her brother, was a flapper, one of the jazz age young whose rebellious reactions to prosperity and change are well documented in American history and literature. Putnam matured in harder times. In 1990 he protested to his sister, reiterating their mother's rule to put family first: "Generosity was part of the fabric of her life."[17]

Putnam's considered reflections provide balance to Fran's views. She liked to tell of her father's status as a dollar-a-year man for the government; Putnam records distinguished tours of duty as a special agent for the Treasury Department and deputy chief of the newly formed FBI in the New England area.

Each of the Flint children achieved some distinction. Bertram was a highly successful salesman; brother Vasmer, (1915–1951), a war hero who survived World War II combat in the Pacific, was killed in the crash of his patrol bomber in Korea; and Putnam, admired by Fran as an inventor and canny businessman, is a generous family man, an inventor, and public benefactor.

Fran's memories are of the dramatic incident. In 1983, she boasted of her secret life to Putnam. At fifteen, she said, under the pen name "Claire Windsor," she rented a post office box in order to submit a "true romance" story: "What a story—a girl so beautiful that she stared in the mirror—fearing to break the hearts of men." She got an acceptance check, but one that was too small for the canoe she had hoped to buy. The *nom-de-plume* came in handy, she said, when the Smithsonian Institution Press delayed in giving her a contract for her book, *Harrier, Hawk of the Marshes*. As Claire Windsor, she wrote a "hot letter" demanding a contract for Fran Hamerstrom. Not surprisingly, she felt she had to add a postscript to Putnam: "This is all true."[18]

In the chapter "I Try to Explain my Parents," in *My Double Life,* written after both her children had gone their own ways, Fran admitted, at last, that she "might well be more tolerant." She could then say that she knew that her parents loved her, and finally she accepted her childhood—where she often felt alone and desperate—as an effective training ground for the very traits that she relied on as an adult: initiative, courage, and independence. Even then, she never seemed to recognize how much she resembled her father in her willfulness and determination; nor did she see that she, like her mother, regularly used positive approaches to difficult situations.

Fran's feelings ran deep. Posey died when Elva was on her way to

Plainfield for her yearly visit. When she arrived, Fran met her in the yard. "Posey died," she said briefly. "I don't want to talk about it." And she went back in the house. They never did talk about it. The strength of Fran's emotions about her youth comes through in an often-reiterated statement, "Both Hammy and I, for different reasons, maintain that never, never did we want to be children again."

In one of Fran's flashes of perception she remarked: "Hammy came from a happy marriage. I came from an unhappy marriage. Our marriage has been happy. What do you make of that!?" Since Hammy never opened the reservoir as she did, one can do no more than speculate. We know that opposites often attract, and that a shared love of the natural world can bridge many differences. The tastes and traits they shared, as well as their clear differences, became more obvious over the course of their long marriage.

Hammy's family history is thin: no family historian fills Putnam Flint's role as the guardian of dates and records. Hammy's younger brother, Davis, provided some information in a 1989 tape of his childhood memories, along with a slim folder of clippings and letters.[19] Photographs of two of his boyhood homes display spacious residences. Fran, who loved telling stories, thought one of the family traditions was significant and often told it to her children. "Thor took a hammer, and cleft a rock. A stream gushed forth: the Hammer-strom, our name," she would intone. "At that time, family holdings among the gentry were passed on to the eldest son. Your great-grandfather, William Hamerstrom, got only a gold watch. Furious, he nailed that watch to the wall over his bed and shipped out to America."[20]

Diligent, resourceful, and independent Hamerstroms were among Hammy's forebears. William, that immigrant grandfather, settled in Galesburg, Illinois, where he ran the blacksmith shops of the Chicago Burlington and Quincy Railroad. He was remembered as "one of the 'freak' men of the town," by his daughter Ruby Hamerstrom Darrow—in her old age a prolific letter writer. She appreciated the fact that her father made good use of the public library, read Voltaire and Tom Paine, and subscribed to a wide variety of the liberal press of the day.[21]

William and his wife had four children. Ruby, the eldest, became the adjunct mother of her three younger brothers (Frederick Nathan, born 25 May 1873, Burt, and William E. Hamerstrom, who at one time served as the quartermaster general of the Army of the Philippines).[22] Ruby got tired of understudying her "poor, distracted" mother. With her thick red pigtail hanging down her back and her small savings in hand, she left Galesburg for Chicago. There she became a writer for the women's page of the *Chicago Evening Post*.[23] When she met the divorced, forty-two-year-old Clarence Darrow, Ruby managed to mute his preference for "free love" and became his second wife: "She straightened him out," commented Davis.

Darrow became very fond of Frederick Nathan Hamerstrom Sr., his wife, Helen, and their two promising sons. He became a significant influence on the intelligent and skeptical young Frederick Jr. How Fran loved to report, "*Clarence Darrow* is his uncle, you know!"

Frederick Sr. headed directly into commerce and began as a "drummer," a traveling salesman, in the Midwest. His territory included Wisconsin. An old photograph of a handsome, dapper young man sitting at a railroad station in a small Wisconsin town establishes the midwestern location. A 1904 letter from his younger brother Burt indicates early lean years for Frederick Nathan Sr. Burt wrote philosophically: "I am sorry but not surprised to read of your hard year. . . . Seems as though you're getting to where you might begin to float and quit sailing. If you will notice, you're tossed about a good deal anyway; row as you will. . . . If the port is Wealth one must ever add more canvas and oars; if it is Happiness one must trust to his barge more than his own efforts."[24]

Hammy's father continued to "add canvas and oars." Early in the twentieth century he moved to the Boston area where he met and married Helen Harper Davis, "lovely Helen," in sister-in-law Ruby's phrase. Helen was a year older than her husband, and was, according to Fran, always embarrassed about that fact.

Helen Harper Davis, born on 14 October 1872 in Philadelphia, was a belle who did not marry until well into her thirties. She had marched in suffragette parades in a Gibson Girl blouse and an enormous hat. She and Frederick Sr. established a series of pleasant homes in pleasant towns, the last of which was Winchester, Massachusetts. She was a lifelong Episcopalian, but her elder son was unimpressed with this practice. Hammy's application form from Harvard asked

Rising entrepreneur Frederick Nathan Hamerstrom Sr., in the early years of the twentieth century.

Helen Harper Davis as a bride, at the turn of the the twentieth century.

about previous employment; he answered, "Sunday school superintendent." Davis judged this unlikely entry to be "a joke."

The engaging reminiscences of Louise Davis, Hammy's maternal
grandmother, bring further themes, those of development and environment, to the story: "My father, Nathan Harper, was . . . the first
fork maker (Hay and Manure) in those parts. . . . About 1836 or 1837
he moved to what was afterward called Harper's Hollow. The mill
must have been a gristmill or flour mill as the house and mill were
built together. . . . The mill was run by water power—a large wooden
trough carried the water to an immense wooden wheel, which was my
childish delight to watch. The house was wonderfully built of stone. It

was a fine old mansion. The water we used . . . came from a spring cut out of solid rock and was icy cold. This spring supplied the water for the houses of the workmen on the place." The columnist quoting her added descriptions of wooded hillsides and a bird-filled dell around a creek clear enough to bathe in. He went on: "But times have changed. . . . The Dam has been filled in and used by the Belfield Country Club as part of their golf course. A quarry has been opened on the west side of the dam, and the refuse from it scattered around. Where the creek flowed a huge sewer has been built, the covers of the manholes being fifteen or twenty feet above the grade of the paths that ran through the hollow. . . . Eighteenth street . . . runs over the sewer and at this point Chew street will cross and will obliterate the remaining buildings in the hollow. The junction of these highways will . . . make of this spot a busy center . . . over the graves of the old mills and streams."[25] In those days, such growth was recorded with mild regret. For Louise Davis's grandson, growth and development would be considerably more displeasing.

Frederick Nathan Hamerstrom Jr. was born on 8 July 1909 in Trenton, New Jersey, shortly before his family moved to Winchester, Massachusetts. His brother, Davis, was born four years later. The following letter from Hammy's ninth year brings young Frederick to life:

Dear Daddy:—
This is the way I took my bath:- I got the water about 9 in. deep. Then I started to wash. I washed my face, hands, neck, ears, legs, and I did my body. I forgot to wash my hier [*sic*]. I scrubbed my knees as hard as I could. The water was not very warm so I did not stay in long. When I was through with my bath the water was black. Then I ran the water out and washed the tub. Then I dried myself with my towel. After I dried myself I put on my pajams and went to bed.
Sincerely Yours Frederick [26]

The letter sounds like Frederick Hamerstrom. It is a bit formal. It is honest. It is methodical: exact observations describe a systematic mode. This lad accepted convention. Scrubbing the tub could have been simple obedience, or an indication of his natural fastidiousness.

The Hamerstroms Sr. were a harmonious family. The boys and their father often made ice cream in a big old freezer on summer

Sundays. "My dad was popular with the neighborhood children when I was in college," said Davis. "They used to come to the back door asking, 'Can your father come out and play?'" Hammy testified, "My father never crossed my mother." The children seem to have had everything: gentle, indulgent parents and grandparents; elaborate toys, sumptuous clothing; a cook; and always, a dog. They had fine furniture and a marvelous touring car from which his mother, in an old photograph, resplendent in black sealskin, smiles radiantly. In Hammy's early boyhood, summer vacations were spent in Cape May, New Jersey. Hammy loved Cape May. A kindly streetcar motorman used to let him pretend he was the conductor and clang the loud bell. And Grandfather Davis, slim and handsome, always took him along on his fishing trips.

Helen Davis Hamerstrom cherished her beautiful homes. She took care that the lovingly arranged family album she sent to the ramshackle farmhouse in Wisconsin in which Hammy and Fran then lived included enlargements of houses and certain rooms of furniture that Putnam remembers as tasteful.

When it came time to choose which of their family's worldly goods the Hamerstroms Jr. wanted, "They took one piece from all those beautiful things! They chose a little desk."[27] Today's Helen Hamerstrom, wife of Hammy's son Alan, shrugged as she spoke: "It is good," she said, " It was our wedding present." Fran explained: "We were influenced early by Bertrand Russell. He put the choice well. 'To have,'" she paused dramatically, 'or to be.'" She liked the phrase and often repeated it.

Young Frederick spent a great deal of time in boyish outdoor pursuits. He caught snakes, braided them together, and threw them in the canal at the back of their house; then he watched them unwind and swim away. One memorable day he and a neighbor's son climbed out of an upper window onto the drainpipe of a roof projection. Discovered by Hammy's distraught mother, who was "more surprised than pleased," as Davis reported, the nine-year-old augmented his guilt by laughing when she spanked him. At nine, he was too big to cry. She reported his effrontery to his father that evening. Davis, not yet in kindergarten, was sent to bed before parental authority was asserted. He stole down the staircase and peered around the arch into the living

Hammy and his mother, on vacation at Cape May, circa 1913.

room. There sat his father in the swan chair, its arms curved over in carved bird heads. Frederick stood before him. "My father wasn't much on spanking, but this was a crisis. He got his fancy cane, put Frederick across his lap, pulled the cane up in a dramatic flourish— and broke the chandelier. It came crashing down—and worse, the cane hit his knee and cracked in two. I'm not sure he hit Fred at all. Everybody was crying. I couldn't stand it—I went back to bed. That was probably the last spanking anybody had at our house. It is one of the first things I remember."

Davis often found his quiet, shy brother reading in a big chair, the apple core he had been eating buried in the cushions. Hammy preferred reading to roughhousing with schoolboys, whose amusements might result in his being pummeled and rolled down the steep school hill into the barberry bushes.

They lived in an era when love for the outdoors came easily and naturally. The two boys spent hours roaming the wild spots near their home. Davis's memory was vivid.

Lillian, the maid, watching young Hammy hard at work with his tractor in 1912—review of what was to be a frequent activity in the Sand Counties.

In 1923, there was a lovely pond out at the back that got this clear black ice that would ring when you threw a stone on it. You could see right through it and see the grasses and goldfish below. Father would skate around it on his clamp skates. One day, when we were out there by ourselves, Fred broke through and had to get out by breaking the ice all the way to shore. He just went along home and changed his clothes.

We often went to an old abandoned quarry. The kingfishers used to make nests up on those high cliff sides. We found a little cave—it was more of a crack than a cave—where Fred went in as far as he could one day. Then he couldn't get back out. I was scared and wanted to go get help, but he said no. It must have been close to an hour, though it seemed much longer, before he worked his way out. We never said anything about all that to anyone.[28]

In high school, Davis Hamerstrom saw his older brother as a dandy. "He and Crosby Kelly—they and their derby hats! They took young ladies dancing." Hammy loved dances, and he loved the music of the time. He built a crystal radio, and when reception was poor, he hooked a wire to the bedspring and strung an aerial out the window to a tree. Eventually he moved up to a splendid, big, ten-tube Super-heterodyne radio.

Hammy was allowed to handle his father's treasured L. C. Smith shotgun and learned to treat it with respect. He became a very careful, accurate shot. Davis watched him skin his first woodchuck for use as skunk bait—skunk pelts sold for $8 in 1928.

Hammy only very rarely mentioned his family. "My father," he once remarked in his deliberate way, "was a man who made and lost several fortunes rescuing companies in trouble and, sometimes, turning them around." But a long letter from Aunt Ruby Darrow after Clarence's death speaks in persuasive detail about possible health concerns and real financial uncertainty late in her brother's career.

[He] had been financial . . . manager of big concerns, but—in with a bad crowd,—the Converse Co.—who kept him abroad . . . establishing new—and more—branches—on which they lived. . . . Clarence [Darrow] prevailed on Frederick to dissociate himself, which he did—then the accident! Preventing getting into some other concern . . . He has not been idle; he has developed a system for sanitation in Mass.—thoroughly approved by governor and others, gone through patent-depts in Washington, so seriously needed that it is said that not a body of water in the state that hasn't a bed of slime. But politicians see the chance to . . . graft from—instead of letting it go into effect at minimum cost to tax-payers . . . but expected to go through—"any time."[29]

The boys were careful of money; a brief fireworks display struck them as a waste of their five painfully saved 1924 dollars. Hammy's lifelong prudence with money started early; his father's boom-and-bust finances may have been an influence, but his summer jobs and the need for saving for college set certain habits. He was a camp counselor at Camp Abnacki on Lake Champlain and later worked at the Vineland Station University Experimental Farm near Hamilton, Ontario. Of that summer he wrote,

This is a great job. . . . Of course I'd hate to be a farm hand all my life . . . [but] just now its the nuts . . . hot sun, good food . . . and sleep. . . . Its hard work, undisguised and unsoftened for the greenhorns."

An old chap sixty-nine years old and I ran the mixer. . . . I let the old duffer do all the light work—filling the water buckets from a barrel beside the machine, and pouring cement powder into the bucket, while I crashed thru with the heavy work . . . Filling the hopper with wet gravel, swinging a forty pound bucket of cement up on top of the dirt, opening the feeder, shoving it in, running around in front to dump in water, then . . . back again for more water, loading in more wet gravel, and rushing around at top speed all day long. . . . I was about bushed when we quit. . . . We put in a cellar, a cement roof, and laid a side-walk in four days. The boss said we were working so hard he let us off fifteen . . . minutes early each day—something which never happened before in all the seven years one chap has been there.[30]

The consideration of the older worker and the diligence indicate a reliable youth—a contrast to Fran. While he was producing a model botany workbook, she was designing disruptive student action at Miss Goodwin's school.[31] While he was a counselor for boys at camp, she was close to being judged a delinquent. Saved, in her account, by a wise judge's explanations of the consequences to others of her breaking windows or stealing warning lanterns, she "became considerate."

However, this didn't happen all at once. Imagine having had the young Fran as a daughter or sister. Putnam's 1936 story, titled "Snake," told of her demand that he stop the car on the Newburyport Turnpike (she drove with him to his private school in order to bring the car back to Milton). The turnpike runs next to a swamp; she took off a silk stocking, ran into the swamp, and "recovered a three-foot long snake which she put into the stocking and tied it firmly." Three months later, at home for a winter vacation, Putnam found that the snake, seeking warmth, had found its way to a compressor in the cellar. She had freed it in his bathroom at home, a room never used by the parents. Fran was then reveling in teaching a crew about fieldwork in Necedah, Wisconsin; her prank seems inconsistent with her declared attitudes toward living things.

Hammy valued scrupulous honesty. Davis cited his father's example. "Dad was a ramrod; he pushed that into us. Indeed, a judge once told me that my father was 'too honest.'" Fran's dramatic approach, on the other hand, worked for effect rather than fidelity to fact. The two of them agreed on the "good breeding," which characterized their respective families. Fran would accept no designation for her lineage but "aristocratic." Both Davis and Ruby Darrow used the words nobility and gentry for their forebears. Putnam recorded his father's status as a Mayflower descendent (Cushing), a descendent of Dr. Charles Chauncey, second president of Harvard, and the fourth great-grandson of Captain Samuel Flint (wounded at the Battle at Lexington and killed at the Battle of Stillwater in the American Revolution). He was also William Rice Carnes's grandson. Carnes was Zachary Taylor's aide-de-camp. Both sets of parents came from stock that fought in the colonial wars.

Hammy was too polite and essentially too democratic to stress rank or status, but Davis's account of a hike along a stretch of a wretchedly rainy Appalachian trail offers a possible corroboration for the family legend. They met up with an old codger at a shelter. When the brothers introduced themselves the old-timer exclaimed, "My family served a judge's family by that name in Norway. It must be the same people! Let me have those wet socks!" He hastened to dry the socks over his fire, and in the morning he went ahead to have the next shelter warm and ready for them. "He couldn't do enough for us." Coincidence? Perhaps. Davis and Hammy would never have claimed any privilege of lineage, but both had courtly manners and aristocratic bearing. The matter might best be referred to another definition: that of E. M. Forster who speaks of: "the aristocracy of the considerate, the sensitive, and the plucky." In that light, aristocrats abound, in both the Hamerstrom and Flint families.

# 3

# Self-Discovery and Love

*But I do not yet know with sufficient clearness what I am, though assured that I am.*

—Rene Descartes

Two young men boarded the Boston & Maine Railroad's Montreal Express from the station in Winchester, Massachusetts, in September 1927. The slight one, Thad Smith[1] carried a suitcase and a tennis racquet and within the day was to become the roommate of the tall, handsome, and raven-haired one—Frederick Hamerstrom. Both of them were graduates of the high school in their pleasant small town; Thad Smith's mother was delighted to further her son's connection with that "lovely family." At White River Junction an old bus awaited them; it took them to Hanover, New Hampshire, and Dartmouth College.

Dartmouth's 1927–28 costs were $860, plus books and incidentals.[2] Hammy hastened to find a job—as waiter, soda clerk, and cigarette salesman for forty cents an hour at "The Wigwam" coffee shop.[3] The next year he was kitchen help at the Atkins Eating Club until it failed. Thereafter he solicited new accounts and collected laundry for Brown's Laundry. "Not particularly pleasant work," he noted dryly. Odd jobs brought fifty cents an hour. Such earnings, he claimed, paid 29 percent of his Dartmouth expenses.[4]

Hammy and Thad shared a second floor suite in Fayerweather Hall, a small three-floor dormitory with thirty rooms. Amenities included a late-night break as the resounding call, "Toast Sides!" brought residents pell-mell downstairs to a vendor in an old truck fitted with a lantern and charcoal grill. Toast with jam or peanut butter, and coffee, tea, or soda cost twenty-five cents.

34

Thad soon pledged Theta Delta Chi fraternity. Self-sufficient Hammy preferred to roam the banks of the Connecticut River, noting the alders above the Ledyard canoe club, exploring the bluff near where Blood Brook comes in and determining that the fishing there was not worthwhile. His life was routine: "I have classes in the morning, and rec in the afternoon for an hour and a half, and then work in the evening. In between times I have to study, and when thats done I have only a little while to sleep."[5]

Letters home repeatedly reassured his parents of his welfare and of the value of his pursuits. Roles seemed reversed—the student heartening the parents. Was he trying to convince a parent, or himself, that he was doing the right thing? He must have expressed some doubts to Clarence Darrow, whose visit in May 1928 was memorable: "Uncle Clarence and Cousin Paul say to come back next year, and also to finish. Uncle Clarence said I was doing damn well (he said damn!) and that I might as well go thru with it. So I guess the money I can't make needn't cause Dad any worry or expense. Its certainly giving me a break—he must think I'm doing well enough to deserve it, or he would'nt offer it."[6]

Darrow might have offered a loan, although Fran insisted that Hammy got no help from him.[7] Hammy may simply have been countering a parental opinion that employment was more practical than college. Thad Smith heard nothing of financial difficulties "until the crash made things hard for everyone." Withal, it seems likely that Dartmouth expenses were a burden for Mr. Hamerstrom Sr.

Darrow's son by his first wife, Paul, an alumnus of Dartmouth, and his daughter-in-law accompanied him on the May visit. Hammy had mixed reactions and reported that Darrow "called on the president Saturday morning and Bill [a friend] and I were allowed to sit in before the hallowed presence—an unusual event!" Still, the freshman could not resist a little condescension: "Typically mid-Western, are'nt they?"

Hammy invited an awed Thad to dinner with Darrow at home. Thad never forgot it: the most famous attorney in the country after the 1925 Scopes trial dazzled him. "I can, even to this day, picture him sitting at the dinner table with that black cigarette in the center of his mouth."

"He always smokes black cigarettes," Hammy told his roommate, "and he always holds them like that, right smack dab at the middle of his lips. He never puts on airs. Dad says that his office in Chicago is just a small, ordinary one."

Mr. Hamerstrom Sr. had arranged for $1,000—a significant sum indeed in the early years of the twentieth century—to await his sister and Darrow on their honeymoon in London, with the modest message, "a bit more voyaging for you two dear ones."[8] These "dear ones" found Mr. Hamerstrom's marriage devoted, and felt that his household in Winchester (the prettiest suburb of Boston, said Ruby) was situated among interesting, cultured people. The connection did not weaken with time. "Never go back on Fred;—see him through," was one of Darrow's last admonitions to Ruby.[9]

Darrow's support for the underdog and unpopular causes, his intellectual vigor and independence, and his clear stance against organized religion appealed to his intelligent and skeptical nephew. A permanent connection had been formed. Clarence and Ruby ignored stuffy conventions. They lived in the top flat of the six-story Hunter Building overlooking Jackson Park, the lake, and the expanding campus of the University of Chicago. To make a library large enough for his thousands of books they simply knocked out the walls of two rooms. They traveled, they had glamour. Some years later, impoverished graduate students Hammy and Fran drove down from Madison to Chicago. They arrived in Oak Park in the afternoon. Aunt Ruby came to the door, opened it carefully, and greeted them in a whisper.

"Your uncle is asleep. But I know he'll want to see you. Come back at 2:00."

"Two in the morning?" stammered an astonished Hammy.

"Yes.

"But you'll be asleep!' Aunt Ruby pursed her lips.

"Just look for a light under the door." At two o'clock, they walked into a party. They feasted on imported delicacies with actors and actresses, European writers, professors and philosophers, listened to extraordinary opinions—all of which made an impression.

The 1927 Connecticut River flood marked Hammy's freshman year at Dartmouth:

Last Thursday I . . . dug mud out of cellars all day—and did we work! The town was simply wrecked—houses torn from their foundations, uprooted trees, roads washed away, and mud—mud everywhere! Cellars filled with it—floors and tables covered with it—fields and roads buried beneath it—mud, hard packed and frozen, caked nearly as hard as cement, to be broken only with the aid of axes. And debris . . . clinging to the branches of trees out of reach overhead—piled in great heaps against houses, stumps, rocks or any obstruction—hanging from fences, piled in huge mounds, tangled and twisted, ruin everywhere. We stood in mud up to our knees in that cellar and shoveled—and shoveled—and finally poured it out with buckets along a bucket line. . . . We stood on Ledyard bridge—one of the few which stood—and watched the stuff rush underneath. Barn doors, fence posts, trees, furniture and all sorts of rubbish went shooting underneath with terrific speed—an awesome sight.[10]

For several days student volunteers who helped missed classes without penalty. Hammy scavenged odd objects (Thad thought them "gross") from the detritus. Farther south a noticeably pretty Smith student named Frances Flint went to the rescue too. "I washed babies!" she later reported.

His fellows saw Hammy as a real student. Thad admired his "facility for remembering the Latin and Greek names, and his talent in the sciences." Actually, Hammy preferred philosophy, psychology, and fine arts. English, once he got past the routine freshman course, was a pleasure; the required math was at the bottom of his list. He did not neglect the truly important skill for the college man during prohibition: finding a source of alcohol. Never mind the rules against liquor in the rooms; Hammy discovered Joe Pilver, a bootlegger who lived in Vermont. Once doctored with bitters and juniper, his product became an almost palatable addition to fruit juice or soda.

Hammy would not miss the dances, happily for him, because in the fall of 1928, he took Fran to a dance. She had come to a Dartmouth house party, but her blind date—a Fayerweather Hall resident—couldn't hold his liquor. He had the presence of mind to call for someone to take his girl to the dance. Hammy, who had heard her laugh and her voice, "low, melodious, unforgettable," was an eager stand-in.

Miss Frances Flint in 1927, as a demure Smith College freshman, found beaus and fashion more compelling than her classes. In her words, "I flunked out."

The cheerful, dapper Frederick Nathan Hamerstrom Jr. that Fran met in his
sophomore year at Dartmouth College, winter 1928.

He intrigued her. In a tuxedo crowd he wore a well-cut tweed suit. "I can't afford a tuxedo," he said matter-of-factly. He explained his name; the oldest Hamerstrom son was always called Hammy. He was "a heavenly dancer." She imagined living with him—or dying without him—and directly upon getting home, put his name in her card file, where it remained. The yellowing, brittle card held his Dartmouth address and a penciled notation in her decided backhand: "Socks 11 1/2; clothes 39 long; shirt 16 neck; 32 1/2 or 33 sleeve." He was 6'2" and weighed 165 pounds. They became secretly engaged on their third date and often quoted their subsequent conversation.

"I asked 'What took you so long?'"

"We Norwegians are deliberate types."

Such repartee delighted her; she remembered his laconic one-liners, stored them in her memory, and began to perfect her role as straight man. A previous attraction paled; the more this sought-after beauty saw of Hammy, the more she wanted him.

He, who had been until then something of a loner, had found what he maintained was the only woman for him. He must be near her. He must see her often. He wanted to be welcome in her home and to announce their engagement. Long letters in his strong backhand arrived (and were sometimes intercepted—in the way of young brothers) at the Milton home. She managed to visit him at every opportunity for the rest of the year. The rules forbade women visitors in the dorms, but they ignored the rules. Thad's father (Dartmouth 1896) on a visit to his son, found her in their suite. Presently the dean, who was Thad's father's classmate, called Hammy in. Years later, in Denver, Fran turned abruptly to Thad, "Do you know that your father almost got Hammy suspended from college?"

"What? What do you mean?" he stammered. Laughing, she explained.

"He never even mentioned it," Thad reported. "Imagine a roommate who would go through something like that and never complain! He was always a one-woman man. I never saw anything like it."

Hammy was in love, and his ardor gave Fran a new and welcome sense of herself; he concentrated on her to a degree I only heard about years later. One day I remarked, "Fran, that wine-red outfit is the perfect color for you!"

The youthful Frederick Nathan Hamerstrom Jr. in the late 1920s. "His hair was as black as a raven's wing," said his Aunt Ruby.

Unexpectedly, Hammy responded. "That's the color of the dress she was wearing the first time I met her."

"Hammy!" I expostulated "You don't mean you can remember something like that for forty-five years!"

"Of course I can," came the quick retort. "She wore it to a dance in Dartmouth." He paused, considered. "It wasn't always easy to see her, you know. I think I can remember every occasion I did. I had to be very careful when I called to arrange it."

"What in the world do you mean?"

"If her Father answered when I rang I simply had to hang up."

"Hang up?"

"Oh, he would have never allowed it! He had a terrible temper."

"And then," Fran interposed, "I had to listen to my father stamping up and down the hall, roaring and swearing. 'Those fools!! They can't run a telephone company! Damned incompetents! I'll have the beastly thing taken out.' Of course," she went on, "I never let on. I'd just go out and call Hammy from another phone. I always had to hide the things I really wanted to do. My childhood made me expert at managing."

Fran thought of a way out of these difficulties. "Why not just transfer to Harvard?" she asked. He saw problems—convincing his parents of the wisdom of such a move, and finances. But he wrote to Harvard directly, stating his determination to transfer, "irregardless of the factor of graduate work." In answer to a demand for a "full and candid answer" to the question of why he wanted to enroll, he wrote one bold, black, line. "I believe Harvard College has more to offer than any other."[11] Candid? Perhaps. Full? Not at all.

To finance what lay ahead, he applied to a luxury hotel in Rye, New Hampshire, hoping that it would pay more than the room, board, and $60 per month he had earned previously at Camp Sangamon in Vermont. Fran's reaction to these plans was swift and surprising. He should dance with her, race boats with her, hunt with her. Instead, he was going away for three months! He didn't care! Her retort was prompt. "And *I* am going to California for six months."

Fran's father had considered his daughter a lightweight and was scornful about her chances for employment. "Somebody might take you on as a governess," he concluded. To her mother he grumbled,

"She'd *better* marry well if she expects to continue dressing as she does." Fran, however, had begun to use her looks. She had discovered that fashionable Boston department stores paid good money to clothes models. She got an interview. Then she fielded the customary question of experience with a haughty statement, "I have been accustomed to wearing beautiful clothes all my life," and she got the job. Facing a summer without her suitor, she simply took off, alone, for California, where beautiful young women had a chance for genuine fame. The move was more than a declaration of independence from a constraining family, more than an attempt to become "so famous that I did not have to hide how odd I was." It was also a test of Fran's and Hammy's feelings for one another.

Aunt Ruby, who admired her handsome nephew's choice of partner, described that trip. "Frances . . . an improvement on jewels, furs, gowns, what-had-they . . . moved westward, waited on tables in restaurants where proprietors begged her not to leave, as she brought business—but, on she went; clerked in department stores, sold things at doors—tried every possible 'different' situation."[12] Meanwhile Hammy and Davis were "glad to get $50 a month plus room and board," at the famous Balsams Resort in northern New Hampshire. Here in spite of the market crash the year before, golf matches were held each Sunday, and the stock ticker and the *New York Times* attracted daily notice. Davis was the vegetable cook and assistant, Hammy the caddie master and laborer.

Impatient to hear Harvard's verdict, Hammy sent the admissions office his summer address on 26 June and on 10 July was admitted to registration as a candidate for the degree of A. B. He had to return a bond and accept the requirement of a reading competency in his choice between Latin, French, and German. He wrote immediately, jubilantly, to Fran, and appropriately to the chairman of the admissions committee:

My dear Mr. Pennypacker—
It gives me great pleasure to learn that I have been admitted to registration at Harvard as a provisional Junior. I wish to thank you again at this time for your kind interest and assistance.
I am, sir,
Respectfully yours,[13]

Flapper Frances Flint took the professional name Frances Flynn and, after the 1929 crash, became a highly paid fashion model at Hixon's, a prestigious Boston department store.

In Fran's mind her Hollywood stay brought, in turn, a chance meeting with Gloria Swanson during a photo session; the star's advice, "You should try out for the movies;" and eventually, a contract offer by Warner Brothers. She rejected it in order to return to Hammy.[14] Whatever the whole truth, Davis's understated account suggests that there may have been an ultimatum from Hammy at the Balsams: "It was quite an eventful summer, and included Fran's visit from California in late August in her green two-seater Essex roadster."

Years later, when I asked Fran her formula for marriage, her startling answer was: "I learned never, ever to mention the name of a young man I had thought I was in love with before I met Hammy." She pulled out one of her early poems. "You may use it," she said.

> When I was a young girl
> Anguish came my way
> I put on a blue velvet gown
> and danced in the moonlight
> until I found my song:
> To walk with grace
> in every place
> down to the gates of hell
> is love.

Use it? Where? "Anguish" and "the gates of hell" hardly fit their exultant accounts of their courtship. Nor did her repeated avowal, "I thought I would die if I couldn't live with him." Might not the poem attest to the difficulty of giving up a potentially glamorous career rather than a previous suitor? She said to me once, confidently, "I could have been as much of a star as Katherine Hepburn." She might well have been: she lived several roles, and played each of them to the hilt.

Hammy, meanwhile, was dealing with actualities. He had confessed to his father that the refreshment stand he and a friend had undertaken had failed; $50 would take care of his final obligations at Dartmouth. Mr. Hamerstrom Sr., on the road looking for employment and understandably worried about money, advised that his son plan to work during his Harvard years. Hammy replied:

Your ideas on working my way thru are a trifle out of date. . . . I really know what I'm talking about . . . the head of the Personnel Bureau . . . tells me to borrow. . . . This thing of appreciating it more by working for it is the bunk, Dad. . . . I tried it last year, and I know. I did'nt say anything about it because I knew that you had enough to discourage you without . . . [my] beefing, but it was rotten. I did'nt get to a single lecture, concert, show, or anything else outside of the college curriculum, and I was often only too glad to take some of my cuts just to sleep. . . . I know that I cannot get a Phi Bete key . . . if I have to lose three fourths of my time working . . . you cannot undertake . . . the debt . . . I'm going to take care if it. . . . And I certainly do not want to let Mother crash thru with her Ford stock. . . . I wont take it, thats all.[15]

His father plaintively annotated this letter and mailed it to his wife. He had only advised that working to finance school made one appreciate education more and had just suggested that a girl who did not want to wait might chafe at a delay due to debt repayment.

Hammy held firm. If he went into debt he would take care of it in his own way. His openness with his father in this matter is impressive and quite unlike Fran's way of dealing with her parents.

He liked Harvard. There was no pressure to join organizations so he could follow his own inclinations. Fran often commented that Hammy "detested that joiner stuff." He made friends quickly: Beauveau Beals, a roommate by virtue of standing next to him in the line for housing assignment, and popular Marshall Stearns, a Hasty Pudding member, history major, and later an authority on American jazz. Nonetheless, Hammy's life centered on Fran.

Years later he wrote, "The less said about undergraduate marks the better!"[16] He slid by some courses, for he couldn't miss tennis, opera, and ballet, balls and tea dances with Fran, and their summer weekends with Davis. "They fished like maniacs all day Sunday, Fran getting all sunburned in the front of the canoe, holding her ears so she wouldn't hear him swearing at the backlashes." He passed a required course in the Bible with one night's study. He did value his English major. His declared interests also included business, but he concentrated on the literature and language courses. He always maintained that they were, "the best possible preparation for almost any career you can imagine."

Hammy and Fran shared some adventures and wild excursions. Once, they were nearly swamped in a boat off Martha's Vineyard. On an island in Boston Harbor where flatcars stood on a track, Hammy pushed some up a small rise and prepared to ride them down to the end

Hunting

of the dock. A red-faced sergeant stopped them. "Do you know what's at the end of that pier?" he bellowed. "Those are cars loaded with live ammunition sitting there!"

Fran said, "Hammy outgrew those sorts of pranks. I never did!"

Their hunting trips were just for the two of them. She had learned hunting practice from her father's instructions to his sons at breakfast. She applied his precepts on secret expeditions and now shared her passion with Hammy. They observed old hunters—on Martha's Vineyard and in Katama Bay in duck season—and learned from the old sportsmen. They were, as Fran wrote in 1989, more aware of bird behavior, "than most modern biologists."[17]

Hud

Hunting proved their compatibility and was exactly the right way to start their life together. Impulsive, Fran needed his steadiness. Quiet and deliberate, Hammy took pleasure in her enthusiasm. Almost immediately they found that they understood each other's subtle signals—a glance, a lifted shoulder. Their hunter's alertness, openness to the total environment, and quick responses brought an immediate intimacy to their interactions. They discovered a mutual wish to live an outdoor life "that included plenty of hunting."

Back in Milton there were tensions. He sometimes stayed at the Flint home—but never in a guest room. To Fran's disgust, they gave him the cook's room on the third floor, empty since the servants had been let go. She managed to be with him. "Friends loaned us their apartments. And it wasn't much of a trick to visit him, up on the third floor; I knew the squeaky stairs to avoid." Perhaps that behavior explains his banishment from the Flint house for a brief period that first fall. Putnam quoted his father's reaction to Hammy's first declaration of his intentions. "You God damned whippersnapper! How d'ye expect to support her? You couldn't even keep her in hats!" After they announced their engagement in November 1930, however, "They allowed him back in the house!"[18]

Hammy's courtesy and self-control, plus his promise that there would be no wedding until after his graduation, brought a resigned assent. He wrote his parents, reassuring as ever: "Mrs. Flint was delighted and blessed us and kissed us both. Mr. Flint was a bit surprized but shook hands on it. . . . we know he is not displeased. I think he wanted a bit of time to think it over before getting enthusiastic. It certainly is a wonderful feeling—as tho I had a half interest in the world! We are both extremely happy and hope that you will be, too."[19]

Hammy had every intention of keeping his promise to both sets of parents, but a policeman caused a change in plans. Early in the spring of 1931, Hammy had cut classes to drive Fran to a modeling job in Florida. They had planned a day of hunting quail in Georgia on the way. An acquaintance and hunting enthusiast, A. Harrison Loache,[20] went along. Hammy had a habit of speeding, and in Georgia a state trooper pulled them over and demanded to see his license. The trooper inspected it slowly and peered into the car. "Where are you going?"

"Florida, sir," replied Hammy.

"What is the purpose of your trip?"

"After hunting here we intend . . . " The trooper interrupted. "Is this young lady a relative?"

Hammy looked him in the eye. "This is my fiancée, Miss Flint."

"And your friend here?" He peered at the youth in the back seat. The Mann Act, forbidding the transportation of a woman across a state line "for immoral purposes," was in force—and enforced.

"I came along for the hunt," drawled Loache.

Fran opened her purse, gave the trooper her most winning smile. "Officer, I have a modeling engagement at an agency in Miami. Let me show you their card." As she rummaged she went on, "My fiancé kindly offered to drive me down, since he planned to hunt here anyway . . . and—oh, here it is." She held out a small, engraved card.

The trooper hesitated, decided. "That's all right, Miss," he muttered. Then, returning the license to Hammy, he said, frowning, "Go on, sir." They drove on in silence. Hammy growled: "I want no more of that!" A justice of the peace in Deland, Florida, on 18 February 1931, married them;[21] they swore their only witness, Loache, to secrecy.

Hammy knew that discovery would have meant his dismissal from Harvard, and Mrs. Flint would have found a secret marriage "unthinkable!" Conscious of the responsibilities of a husband, he began to focus on a career. One that involved hunting might be considered irresponsible. He recalled their naiveté: "Fran and I, planning our lives together, knew that we wanted an outdoor life in some way tied to wildlife (even the term was nonexistent—the expression was "wild life") and hunting. We had no faintest notion as to how to arrange it, and Harvard's placement advisers were equally ignorant. We decided to set up a shooting preserve in South Carolina. The wisdom of the early 30's dictated that the way to have lots of game was to raise it in game farms and turn it loose. It behooved us to learn how."[22]

Fran was enthusiastic. Their friends would surely come and could pay handsomely. "Think of it, Hammy! We can hunt all we want!" No mundane matters of land purchase and management clouded her view. Hammy knew better: "We'll have to find a place where we can learn how to raise pheasants."

The Game Conservation Institute, in Clinton, New Jersey, a trade school operated by More Game Birds of America, was such a place. It was late to apply; Hammy, with finals two weeks away, simply had to study. Fran didn't hesitate. She jumped in her car and drove to Clinton where she was courteously greeted, and firmly discouraged, by the gray-haired director. Fran pressed their case; the director passed the buck—she could, of course, appeal to the Board, which was meeting in New York that very day. She drove the eighty miles to the meeting and succeeding in talking her way in. The gentlemen who heard her request were firm at first: she couldn't do the physical work . . . the men would flirt with her . . . she would be a distraction. "There is no problem with your fiancé, he certainly qualifies. But we simply cannot admit you." Seeing her stricken look, they tried to soften the blow. "Were you married, we might have considered it."

"But I will be married when school starts!" she declared, and the directors admitted both of them.[23]

Meanwhile, Mrs. Flint was arranging the June wedding. It must be the expected thing, formal and elaborate. A letter from Fran's favorite relative, her maternal Uncle Billy Chase, supplies a view of what a marriage partner should be for her. Uncle Billy's "Peppering of Advice" came from Paris, on his way to Monte Carlo.[24] "Dear" Uncle Billy lived with his mother, Grandmother Chase, at Chesham. A man of the world, he had sometimes secretly provided cigarettes and chocolate for his niece when she visited there. He, attempting a light touch, advised: "Regarding . . . matrimony . . . be sure that your young bridegroom *in spe* [sic] has ample earning capacity, can play in his play-time and work in his work-time, and he should be able to earn at least $1000 a month, and . . . to play golf, tennis, swim and all the outdoor sports that you may enjoy with him." She should not hurry: " . . . you have ten years at least to wait and make an appropriate selection."

Mrs. Flint planned a garden wedding. ("The roses are lovely in June.") Brothers Bertram, Vasmer, and Putnam manicured the lawns and the flower and shrubbery beds. It rained on 10 June 1931—a steady, day-long rain. Mrs. Flint, who had no idea that the elaborate ceremony was legally a charade, was relieved that Fran did not complain. Fran and Hammy knew that all they had to do was to enjoy their party and act their parts, which they did, with zest.

Fran wore a gown tailored from her mother's wedding dress of satin, tulle, and Viennese lace, its imposing train topped by a great-grandmother's rose point lace veil designed to sweep out in a deep curve on the carpeted stairs. Guests and groomsmen crowded around the graceful staircase in the Flint's stately home. Stockbrokers, dowagers, tycoons, and an array of the young smart set craned their necks as twelve smiling debutantes in coral-colored lace gowns moved slowly down the broad, curving staircase. Some wondered about the handsome groom. Fran heard one guest tell another, "My dear, she's throwing herself away! He's a Harvard graduate, but has no prospects. His father is a *Midwesterner*, you know."

All eyes focused on the broad landing at the head of the stairs. The slim young woman poised there was breathtaking indeed. But her head was turned toward her unsmiling father, her eyes fixed on his frowning, intense face.

"Frances!" he hissed. "Remember! There's still time to back out."

Years later I asked Fran, "Whatever did you say?"

"I didn't say anything. I just grabbed him by the arm and propelled him down the stairs."

Flawless in bearing and appearance, Fran moved slowly down the stairs, head held high, chin tilted. She greeted friends and favorites with poised nods, an occasional smile. She looked directly at her groom in the archway below and gave him a knowing glance. He, with tuxedoed groomsmen, Davis among them, behind him and his best man Marshall Stearns at his side, appeared serious, composed, and assured. Inside, he felt relief and jubilation. "By God!" he thought, "We've carried it off!" Soon there would be no more need for subterfuge.

Fran and Hammy's elaborate and socially important wedding in June 1931 masked the civil ceremony of several months before.

Two years of growing intimacy with Fran made him confident of future happiness. They would get by, for the time, on $25 a week. Eventually his diligence, concentration, and ability would bring long-term security. Hammy knew himself: he could accomplish "a great amount of work in one concentrated effort." His intelligence, he realized, was "well above average." His personality was pleasing, though not dominating. "I am socially adaptable, make a good impression, but am essentially anti-social by nature," he wrote.[25] The fall term at the Game Conservation Institute lay ahead, his plans for the summer seemed sound. As Fran approached, his face relaxed, and he smiled.

After vows, a long reception line, toasts, dancing, and merriment, the couple escaped at a run, down the porch steps and through the rain into her car. Bertram and Putnam threw tapioca—not rice, which they could not find—at the departing couple. "It turned to pudding in my hair," said Fran

Mrs. Flint had very mixed feelings. What would become of her daughter? Freddy seemed very sweet, really, but the graces that Fräuta had labored to instill—the languages, good study habits, deportment, carriage, the proper seat on a horse, skills with crochet hook and needle—what good would they be in the life ahead? Her serene bearing masked anxiety. Brother Putnam was relieved. "I knew all along what was going on," he said, "though they thought they were so discreet. I felt much better when I heard, at their fiftieth wedding celebration, that the truth was they had been married all along."

Another truth was that the senior Flint and Hamerstrom families became friends. Putnam saw his parents getting along better with the senior Hamerstroms than with their daughter and her husband, who seemed to want to get—and stay—away from Milton.[26] Helen Hamerstrom, a gracious hostess, presided at the dinners they shared, occasions that Putnam memorialized with a succinct, "Great pies!"

Fran and Hammy, after years of time apart and separate lives, embarked on their new path. They were united in purpose and ready for adventure. Nothing could go wrong, and nothing did, in spite of family forebodings.

# 4

# Students, Teachers, and New Horizons

*. . . the experience of love will be reshaped . . . into a relation . . . of*
*one human being to another, no longer of man to woman. And this*
*more human love (that will fulfill itself, infinitely considerate and*
*good and clear in binding and releasing), will resemble . . . this, that*
*two solitudes protect and greet each other.*

—Rainer Maria Rilke

Two entranced young people sit side by side before an old barn, gaz-
ing into each other's eyes. His wavy black hair rises above a broad forehead, his chin is down, his hands clasped firmly between his knees. She
tilts her classic, smiling profile toward him, an expression of teasing
yet triumphant joy on her face. Here are the young Hamerstroms, absorbed in their own private world. They never left it.

That summer of 1931 their assets were vigor, enthusiasm, a purpose—and the monthly check from Fran's mother. They needed to earn,
economize, and save enough for fall expenses. Hammy, with successful camp counselor experience to build on, had suggested a tu-
toring service. They went to students' homes and for $2.50 an hour,
a princely sum then, met client needs. Hammy handled English and
math; Fran, French and German.[1] They did well: the next summer, she
raised her fee by fifty cents an hour, and Hammy became a tutor-
companion.

Family connections brought them the use of an empty stable on an
estate near Woods Hole to use as a base. The bathroom was in the cellar of the house next door; their bed was an army cot in the harness

room. (When their mothers came to visit, they housed them in the box stall.)

They returned to student status that fall—at the Game Conservation Institute—an early experiment in what was then seen as scientific game management. Founded in 1928 by the Game Conservation Society, it was run by an English gamekeeper who concentrated on raising birds, ducks, and quail among others, but mostly pheasants. Its staff wrote and published books such as *Game Birds: How to Make Them Pay on Your Farm* and *A Pheasant Breeding Manual: A Simple, Fully Illustrated Description of Methods That Have Proved Successful in Rearing Pheasants.* The school's original letterhead, used by Fran, read "Experimental Farm and Game School"; funds were augmented by guiding paying hunters to bag pheasants. It was never financially sound, in spite of being taken over by More Game Birds of America, whose sponsor, publisher Joseph Knapp, kept it alive until 1935. In

 1937, More Game Birds of America became Ducks Unlimited.[2]

The young Hamerstroms arrived in Clinton on a benign September day. Hammy, in well-cut plus fours, argyle socks, brown oxfords, and a pinstripe shirt, was dubbed, "the coed's husband," by the shirtless, unshaven, rowdy bunch of men that greeted them. He remained unperturbed. Fran promptly saw that she must win acceptance. To show that crew that she was more coordinated, more daring, and more capable than any one of them, she trained by lifting heavy objects in secret, until she was able to heft huge feed sacks with the best of them. She joined in vermin hunts, catching and dispatching rats with her bare hands as the men harried them toward her. No one knew that she and brother Bertram had perfected the technique in childhood forays in the back pantry of the Milton house.[3]

The school had provided them with a wreck of a farmhouse in lieu of dormitory space and enrolled Fran without question in all classes. The house had a stove, a sink, and a table on which they could host weekend dinners. Students contributed pheasants stolen from the pens for these occasions. Fran cooked them and always finished the feast with her home-baked pies.[4] That combination of bravado and hospitality would always be part of the Hamerstrom repertoire

Fran took to her new roles—wife, student, and hostess—with zest. Determined to please Hammy, she planted pansies and morning glories around their home. If Hammy disliked one of her hats, she put it away. She never complained about the ramshackle house, the foul (her word) institutional food, the impetigo that afflicted them and other students, or episodes of food poisoning. That first fall, Fran's father came from Winchester for an unfortunately timed visit. The outhouse had just blown over. She greeted him, in his formal suit—a vest with white piping, a gray silk tie fastened with a diamond pin, and his gold watch chain draped across his trouser front—with charm, smiles, and a good meal. Mr. Flint inspected the house from top to bottom, but there was no masking the fact that standing in the cellar you could see daylight through the floor. "He looked terribly upset. He turned to Hammy and said, 'A fine place to bring a young girl to! [She was twenty-five.] You can't go on living like this.' Hammy replied truthfully, 'Well, Dad, this is how things are in Clinton, New Jersey!'"[5]

Poor Mr. Flint! A gentleman hunter, he like most citizens of the time, thought animals were either to be hunted or possessed as dependent pets. Like others, he deplored the paucity of game "nowadays" and considered overeager hunters to be "game hogs." But of conservation, much less a career in it, he had no inkling.

Raising pheasant chicks to maturity and managing their harvest for a profit turned out to be neither attractive nor very scientific. Hammy soon saw that he must find a more fitting way to work in his much loved outdoors. He disliked his assignment to guide certain unruly guests, with whom he felt unsafe. Fran was outraged by the behavior of "short dark-skinned men who spoke broken English." They could not shoot, and many disregarded the most basic hunting conventions. One bunch, returning empty-handed from the field, insisted that live pheasants be stowed in bags and hung on the fence as targets. Blasting away, they got their pheasants. In a comment that betrays a certain condescension, she wrote years later, "It has always been a mystery to me that we guided parties of immigrants or near-immigrants, rather than hunters from well-established old American families."[6]

Hammy tried other avenues. He wrote a story for *Boy's Life*; it was rejected, but Fran kept the manuscript. The predictable plot involved a coward, a test (the devil's leap, on a ski hill), and a courageous rescue. Its themes—courage, determination, and perseverance—suggest its author's character: these were the traits that manifested over and over in the course of his career. No longer an amateur and not eager to become an entrepreneur, he searched for a way to become a genuine conservationist. The Seventeenth American Game Conference in New York in December 1931 unexpectedly brought him the desired vision. It was a profound experience:

[There] . . . came the almost legendary heads of state and provincial game departments, some US Biological Survey men, and even a thin smattering of university people—a rich mixture of potential job-givers . . . with considerable awe and trepidation, we came into the presence of that august body. We were appalled. To our naive and intolerant eyes, this collection of national leaders seemed a rather ordinary lot, discussing matters that would have been appropriate to a salesmen's convention—how to raise money, how to influence some political action or other, how to promote this or that special interest. Then, suddenly, a slender, deeply tanned man stilled the room as he approached the podium with speed and elegance. And I heard a voice that I can still remember, although I remember none of the words. I heard an inspiring speaker, an idealist as well as a practical man, who plainly—unlike his predecessors— knew exactly what he wanted to say and how best to say it. This man . . . towered above the others in his personality, in his grasp of the subject and in his ability to communicate. And I knew that—remote as the possibility was—if it could ever be in my power, I would do my best to learn from this man who had set me afire.[7]

Hammy was not given to overstatement; that last expressive sentence is a tribute to his future mentor and friend, and Fran's scientific sponsor, Aldo Leopold—the most important influence in their lives.

Leopold was speaking at what was to be the constitutional convention of American conservation. He defended his belief in the democratic American system of public hunting rather than the European practice of private ownership, captive breeding, and hunting for pay. His plan called for cooperation between sportsmen and conservationist; all must "recognize conservation as one integral whole, of which game restoration is only a part. In predator control and other activities where game management conflicts in part with other wild life, sportsmen must join with nature-lovers in seeking and accepting the findings of impartial research."[8] The Conference accepted his committee's report, and his policy held, essentially unrevised, until 1973, the year the

1931 – 1973

after Hamerstroms retired from the Wisconsin Department of Natural Resources (DNR).

This, then, was the couples' goal: "impartial research." That was it; research would lead them to the new profession of Game Management. It meant a specialized graduate program for him, and a return to college for Fran. She was ready. "I told Hammy I wanted to take courses in veterinary medicine. He said, 'Good idea! Maybe you can.'" Inspired and excited, they returned to Clinton, with new purposes and renewed optimism.

They had found their calling. It was to be in Hammy's disesteemed Midwest, where they found another mentor, one who had grown up in South Dakota, a truly remote and wild place.

In the early years of this century, South Dakota was a sparsely settled region dotted with marshes and sloughs overflowing with water birds. A farm boy growing up found magic in the lonely places. Paul Errington, lamed by early polio and living with his mother and stepfather in Brookings, responded to the landscape. He watched beetles, butterflies, toads, grubs, earthworms, snakes, and garter snakes in the "little wilderness" off the alley behind his home, and he roamed the shores of Lake Tetonkaha on his Norwegian grandfather's farm.

Errington's stepfather taught him to shoot and helped him buy his first traps, which he learned to use through patient trial and error. Skunk and muskrat were abundant, the occasional mink a bonanza. His solitary expeditions through the still, frosty mornings gave him the desire to be a wilderness trapper. Evenings at home, huddled close to the woodstove, he read the books of Ernest Thompson Seton, responding to the gritty realism of his portraits of animal life and the images of woodsmen's skills.

Parental pressure (which he resisted bitterly, by his shamefaced later confession) steered him toward college and a teaching career; but fur paid his way through South Dakota State. His love of hunting led him to Madison and a Ph.D. in zoology from the University of Wisconsin in 1932.[9] He went directly to Iowa State College to head the first Cooperative Wildlife Research Unit in the United States. Iowa at that time still had remnants of the great prairie, "a vast grassland from the Missouri River to the Mississippi River" that was "once-vast and variable natural landscape."[10]

In the fall of 1932, Hammy went to a convention of game commissioners in Baltimore to locate a job. The Iowa state game warden announced a position with Errington's new three-year game survey. The salary was $1,200–$1,500 a year, with perhaps a house provided. Hammy applied directly, in September 1932.

As he hastened to explain to his father, the pay was secondary. The job was a stepping-stone to "almost anything in the field." He added urgently, "Do you know anyone connected to Iowa State College, or any senators or representatives from the state?" He enclosed a draft of his eager letter of application: "I should like to be able to convince you of my earnest desire to get this job. It is the one which I most want. I am determined to work in this field, and I feel certain that this is the type of position for which I am best fitted. I have the education, and the desire to work."[11]

Soon, with his customary positive spin, he reported a "very encouraging letter from Dr. Errington."[12] He and Fran drove to Ames in their run-down car, stopping in Chicago to see the Darrows. The rest of the way they ate roasting ears and roadkills and slept in roadside fields.

Fran maintained that luck alone, "brought us to study with Errington, and then with Leopold." In her version, they landed at Errington's, unexpected. Hammy's account fills in interesting particulars: "The reply said a number of discouraging things, but it never conclusively said no. So we packed up and presented ourselves on Professor Paul Errington's doorstep. Our old car broke down as we entered Ames, leaving us with neither means to get back nor money to repair it. Learning that we had been sleeping in the fields to save money, Professor Errington . . . put us up in his own bed, and slept on the couch."[13]

Errington hired Hammy over 152 other applicants at a salary of $90 a month. Classes had already started, and Hammy began work immediately. Fran found a job at student wages. To her astonishment, she was offered thirty-five cents a *day*, not an hour as she had expected. She gulped, and took it anyway. They found an "ugly house" on Wood Pond Street and furnished it by haunting auctions and secondhand shops in Des Moines since shopping in Ames was too expensive. Fran wrote her mother-in-law in the formal style she maintained for years:

Dear Mrs. Hamerstrom,
We had less trouble in getting what we wanted in the way of furniture than we
had anticipated. . . . List of furniture in good condition

| | |
|---|---|
| dining room set (4 chairs) | 15.00 |
| buffet | 3.50 |
| bench for kitchen | .50 |
| dressing table stool | .50 |
| stove | 15.00 |
| fiddle back rocker | 5.00 |
| desk for Frederick (Library table) | 1.00 |
| desk chair | 1.50 |
| Other furniture | |
| a sofa of exquisite shape and work<br>    manship | 7.50 |
| (I shall have to recover it) | |
| a handmade desk for me | 1.50 |
| (I shall have to refinish the wood) | |

a chest of drawers for 1.50 and one for .60 which will be cream with a pale
blue stripe—the colours in our bedroom.[14]

Hammy's field work began after an early snowfall made tracking
possible; course work came with his January acceptance into graduate
school. His program required forty-two semester hours of zoology
plus twenty-five hours of research in that subject; twenty-one hours
of botany, (his minor), chemistry through graduate-level organic, and
fifteen hours of veterinary science. Fran was admitted on probation
(with a note indicating that her IQ of "B[inet?] 134" removed that defi-
ciency). She had left Milton Academy with a certificate of attendance
and had flunked out of Smith. "I simply didn't go to the two courses
that bored me." The university allowed her credit for eighteen previous hours of German and counted some of the game school courses as
zoology requirements. She dropped math and chemistry but did not
shrink from botany, zoology, and veterinary science. As always, she
made time for what she liked: horticulture (garden flowers), French
literature, and painting.[15] "If only I could really paint!" she exclaimed
just before describing the colors of the exhausting and inconvenient

The "ugly house" on Wood Pond Street, Ames, Iowa, was the Hamerstrom home from late 1932 until 1935.

dust storms as "frightfully beautiful."[16] Her poetry won praise from an English professor. Best of all, she was able to time her classes so that she could go on field assignments with Hammy.

Hammy was not pleased when Errington advised courses in botany. "You'll have to be able to identify plants. We want to know what these birds eat, and where they nest." Fran saw his hesitation, joined him in a course titled "Spring and Summer Flora," and encouraged him. "He simply needed a feel for plant families. But as he began to study ring-neck pheasants and their habitat, he became an excellent botanist. Distinguishing those grasses—it's very difficult, you know." Mrs. Errington remembered Hammy spending a great deal of time with Ada Hayden, a botanist with a particular interest in prairies. The remains of their study area, then a vast expanse, now bounded and restored, has been given her name, the Ada Hayden prairie.[17]

Errington thought highly of his bright assistant; he gave him a copy of Leopold's *Game Management* on its publication in July 1933. He inscribed it: "To Frederick Nathan Hamerstrom Jr. with the benedictions of Paul L. Errington." His confidence was not misplaced; Hammy made the scholastic honorary Phi Kappa Phi the next year. Errington liked Fran, too. "Men usually liked Fran," observed Errington's wife, Carolyn, "but he liked her because of her devotion to conservation."[18]

Errington's interest in predation meant the painstaking gathering of data. For example, what did owls, then considered vermin and as harmful as hawks, eat? Did they really prey upon useful or favored species? Answering involved collecting over five thousand owl pellets and analyzing their contents. To simplify collection, the young researchers tethered the young owls on slopes near their nests. Six years of painstaking identification of bones, feathers, chitin, and such remnants ensued. They identified creatures ranging from large insects to skunks, opossums, and even geese and turkeys.[19] It was slow work. Hammy and Fran found the technique especially useful years later during her harrier project.[20]

Taking daily accurate notes of patient observations became habit. Skills and techniques grew. Fran described how, seated on separate low prairie hillsides, they would watch a harrier return repeatedly to the same spot in the marsh below. "Then we'd wave to each other, and walk in, to meet at the intersection of their walking lines—and, the nest!"[21]

They coupled learning with pleasure on their vacation time, and that practice endured. Errington approved their first excursion—to the Black Hills of North Dakota. Hammy didn't mention their intended return route through Vancouver, Washington, and Oregon, land of the Sooty Grouse (now usually called the Blue Grouse). Camping was all that they could then afford. They loved it; they camped into their eighth decade, long after they had to, thus surprising even hardened outdoorsmen.[22]

A 1934 letter to the Darrows enclosed a copy of Hammy's first paper, with a deprecating, but perceptive, note. "I am sending . . . the first fruit of my scientific labors, which may have more value to you as a curio than as reading matter. I feel quite proud, and very, very scientific, although I am only junior author." He announced a proposed visit in December and admitted a certain reserve. "I wish I could tell you how much we really enjoy our visits with you—we anglo saxons are a queer

Professor Errington and some of his graduate students at Ruthven, then the site of the University of Iowa's field station, July 4, 1933. *Standing, from left:* Chuck Friley, Fran, Hammy, and Paul Errington. *Kneeling:* Logan Bennet and Andy Amman.

lot, at best, in the matter of expressing such things, and I find myself less able than most to really break down and open up. Not like Dad and Dave in that respect, eh?"[23]

In spite of that reticence, they started building a network of close friends. Of Paul and Carolyn Errington he said, "His wife is an awfully nice girl, and a lady. We feel very pleased, for he is such an unsuspecting cuss that he might easily have been roped in by some little golddigger. Carolyn comes from Paul's home town in South Dakota."[24]

A letter to his mother told of a job offer with the Soil Erosion Survey, an offer he kept quiet as "Dr. Drake, head of the zoo dept here, would probably make me take it."[25]

Drake was an uncompromising academic. Five years later when Hammy wrote from Madison announcing that he had passed prelims, he was more explicit. "It is really a tremendous relief. . . . This exam has been in the back of my mind for five years, since the head of the Department at Iowa (not Errington) told me that no one who had not

Uncle Clarence Darrow, in Chicago. This snapshot probably accompanied the Darrows' 1932 Christmas greeting to his nephew Hammy.

majored in the biological sciences as an undergraduate could hope to get a Ph.D. in it."[26] Not meeting Drake's standards seems another stroke of good luck. Hammy would not have earned one of the few Leopold-guided doctoral degrees had he done so.

Errington, a gifted and prolific writer, demonstrated to an apt pupil the language and conventions of science. Hammy's love of accuracy and practice of withholding judgment served him well. His publication list grew; a paper on bobwhite authored with Errington won Wildlife Society recognition. In 1936, his master's thesis was published.[27]

His master's exam was that year too. An elated letter of June 1936, describes a situation that many modern graduate students would like to replicate:

The exam is out of the way—the best master's exam which Paul ever heard— the thesis has been accepted by the library, graduation is over. . . . There is now field work to be done and a perfectly tremendous paper to work on—one which will be a whizzer. I am junior author under Paul who is doing most of the headwork and to whom the credit is due . . . I wouldn't change my lot for anything. . . . prospects are very bright. I have just turned down, for the third time, a $2600 a year job in the Biological Survey, as supervisor of New England forest C. C. C. camps. Mr. Gabrielson, the assistant chief of the wildlife research division . . . has promised me a job anytime I want one during the next six months . . . the man in charge of the game work for the Soil Erosion Service also wants me . . . and Leopold—the biggest man in our field—has asked me to talk things over with him before I go anywhere else. . . . Altogether, the immediate future, at least, seems very bright. . . . Fran . . . was awarded the prize of the Women's honorary society for being the graduating woman of the year who has done the best undergraduate research and who shows greatest promise for graduate research.

We have met and been accepted as friends by the Leopolds . . . a signal honor. Mr. Leopold has asked us to call him Aldo, and we have visited them at their shack on the Wisconsin river—an invitation not extended to many people. In fact, we have been sworn to silence as to the location of the camp. . . . We are having the time of our lives, and would'nt miss this for anything. Paul now considers me his prize assistant, and is a wonderful fellow to work under. He has taught me more, since we have been working under him, than I ever knew existed, and it is a rare pleasure to have him as a superior. He is a prince.[28]

They had come far from the pheasant pens of Clinton, New Jersey. The months of fieldwork, the marshes, the sloughs around the university field station at Ruthven, some two hundred miles north in the western side of the state, the Hayden Prairie even farther north on the east side, and other biologically interesting areas broadened and deepened their outlook.

Hammy still expressed occasional negatives about the Midwest, some directed toward unnamed people in Ames, ("may the good lord save me from them.") His use of the term "midwestern" brought him a parental correction: his father, after all, came from Illinois. Hammy took the point. He apologized that he had made his father unhappy. "The people whom I disliked are not at all like [you] or Uncle Clarence and Aunt Ruby . . . there are many things and some people here which I like very much."[29]

Fran was as firm about following her own interests as she was with being Hammy's valued partner. Aunt Ruby described the pair:

The two became quite a curiosity in that locality . . . pioneers of their kind of work . . . so that people came from far and near to view them and their counters

Aldo Leopold's shack. After meeting Leopold in 1934 or 1935, Hammy and Fran were honored by occasional invitations to join the Leopolds there. Leopold designed the benches in the foreground.

and cases arranged in separate rooms,—for altho' they were not expected to bring forth anything but birds and other wild animal life specimens, Fran developed (through all that they found along their course over their field, besides what they were in search of) a side-interest in insect life and specimens of all sorts—necessitating a room of her own for her line, to which people constantly came to see THE BUGWOMAN — AND — THE BIRD MAN—as they were known to people there, astonished that such life existed in their very environment, of which they had known nothing in particular until then.[30]

Their new life, however, distanced them more than geographically from their families. A report came from the Flints that the senior Hamerstroms were "living in poverty." They decided they must help. Hammy fixed on $5 from his Research Fellow's stipend. Then, at his father's urging, they undertook the demanding drive East for the holidays. Two long, cold days later they arrived at a pleasant two-and-a-half story house. They hesitated.

"Check the address Fran." She pulled a letter out of her purse, held it to the light. "It's the right number." They rang the bell, and a maid in a white apron opened it.

"Is Mrs. Hamerstrom in?"

"Oh, you must be Mr. Frederick! She's expecting you!" Inside, they found fresh flowers on the table. Poverty is a relative concept; the gulf between their choices and those of their families widened. New connections began to supplant family obligations, and when Hammy was offered an opportunity to practice their new expertise in a location within reach of Madison, they jumped at it.

These were foundational years for their partnership. Fran lessened the distance in knowledge between her husband and herself without abandoning her interests in insects, art, and writing, for example. United, but always discrete, they progressed on a path that took them, together, away from the worlds of their childhoods into a more elemental and Spartan life. Their new friends shared their priorities along with an outlook quite foreign to many at that time. To observers Fran may have appeared docile and even subservient, but her husband liked the competitiveness and verve with which she shaped her place in the world they intended to effect.

Their parents were as supportive as they knew how to be. Mr. Flint built them a beautiful little trailer, which Mr. Hamerstrom pulled out and helped them pack for their move.[31] Somehow, they acquired a new car, a spiffy tan Essex roadster. After a final summer at Ruthven, degrees in hand, they headed for Necedah, Wisconsin. It was not only midwestern, it was backwoods midwestern, and the stay there furthered their education in unexpected ways.

# 5

# Conservation Beginnings in a Midwestern Appalachia

*Before Roosevelt, the Federal Government hardly touched your life. . . . Now . . . it came right down to Main Street. Half of them loved it, half of them hated it. . . . But they were delighted to have those green relief checks cashed in their cash registers. They'd have been out of business if it had not been for them. They were cursing Roosevelt for the intrusion into their lives. At the same time, they were living off it. Main Street still has this fix.*
— Ed Paulsen, speaking in Studs Terkel's *Hard Times*

On a brilliant early autumn day in 1935, loiterers on the unpaved streets of the shabby village of Necedah, Wisconsin suddenly came to attention.[1] A jaunty tan Essex roadster was moving slowly down the somnolent street, its open rumble seat piled high with boxes topped by a roped-in plant press and a classic Windsor chair. Behind it, the car pulled a small trailer, also well loaded. Lanky men, their eyes shaded under battered caps and fedoras, watched the vehicle pull over. The driver, a tall, immaculately dressed young man with dark hair and a pleasant smile, opened his door and asked a bystander for directions to the Resettlement Administration office.

The loiterer scratched his head and allowed as how he had never heard of it. Hammy explained that people there were studying muskrats, beaver, and ducks in the area.

"That'll be the relief office. Go on down a block or two and you'll see it; on the right. Used to be the bank."

Heads swiveling, the onlookers watched the driver park, get out, and open the door for a young woman in heels and a tweed suit. Holding her arm, he escorted her into the office.

The diagnosis was immediate and lasting. "City folks." Thus the elegant young Frederick and Frances Hamerstrom were branded in a drought-plagued pocket of the Midwest. Ironically, these blue bloods were as much children of the Great Depression as the impoverished people they had come to live among. They were prime examples of the wide disparity of Depression effects in the regionally and ethnically varied, class-conscious America of the 1930s. They chose, as did later rebels—the children of privilege in the 1960s—to reject many of their parents' values. But unlike many of those later counterparts, they still believed in the worth of an aristocratic breeding and the value of the social graces.

Farm folk, by contrast, had experienced the Depression years as a time of unrelieved deprivation, hardship, and fear. Their hard times started earlier and, in the sand counties of central Wisconsin, lasted longer. A nationwide agricultural depression had begun in the early 1920s and had deepened distressingly by the time of Roosevelt's election. By then, with credit dried up and debt loads already high, farmers were vulnerable—not only to the normal ups-and-downs of markets and the weather but also to larger problems: drought, crop failure related to depleted soil, and takeover by the new large-scale enterprises able to buy the machinery that permitted economies of scale. Five million farm folk—roughly one million families—were on relief in 1933; and many of the slightly more successful still lived in desperate poverty.[2]

The various states did what they could with loans of livestock, improved seed, and fertilizer for those able to follow the recommendations that would rebuild soil and increase yields. It worked for some. Others received sickly stock, late seed, or inefficient treatment by government agents.[3] Those unfortunates now owed money on the unfulfilled promise of the efforts to aid them.

In some places, a long cycle of farm depletion had come to a point of no return. For them, Roosevelt's New Deal developed a bolder program. It aimed to do away with intractable rural poverty by moving farmers from unsuitable land to more promising situations. Biologists would study acres thus emptied; if they proved suitable, they would become flourishing reserves, teeming with game. This was the ambitious task of the Resettlement Administration,[4] established by Franklin Roosevelt's Executive Order 7027 of 30 April 1935, with the visionary Rexford G. Tugwell as the first administrator.

The 164 projects it funded nationwide were a small part of a wider effort: the National Industrial Recovery Act, the Civilian Conservation Corps, the Works Progress Administration, the Federal Deposit Insurance Corporation and the Emergency Agricultural act—instruments by which the New Deal hoped to mend the joblessness, the poverty, and the hopelessness of the times.[5] Wisconsin was part of Region Two, which was overseen by R. I. Nowell, an agricultural economist from the University of Minnesota. The agency, beset by organizational and jurisdictional problems, lasted only until 1937 when it was absorbed into the Farm and Home Administration.[6]

The Central Wisconsin Project consisted of two sites: one west and north of Necedah, and another around Black River Falls. The Winnebago, now the Ho-Chunk Nation, had held that land until 1837, the year when the Chippewa, Sioux, and Winnebago ceded most of northern and western Wisconsin, and the treaty of Poygan Lake opened up the vast area to settlers. It was forested then, with tamarack and spruce in the swamps, pine and oak on the varied upland soils. Bog meadows and marshes abounded, wildlife was diverse but not copious. Early settlers logged off the good timber and established scattered farms, creating new clearings that supported a brief period of wildlife profusion. Michael Goc described a little known time of hardship in the Necedah area: two hundred miles of drainage ditches and a land boomlet in the first fifteen years of the twentieth century resulted in dry marshes subject to frequent and repeated fires that consumed the underlying peat. Poplar and brush invaded the devastated peat; farmers found that the combination of low fertility and the frequent early frost was deadly. Unable to pay high drainage assessments, many simply walked away from their farms.[7]

County boards, facing tax bases lowered by farm abandonment, agreed with the Wisconsin Conservation Department that Necedah, where two-thirds of the total area was in foreclosure or had reverted to the county for unpaid taxes, was exactly the kind of place better suited to wildlife than to people. Local governments had no funds to maintain schools and roads for the scattered families of the region, to say nothing of relief payments for the poor. Individuals yearned for the wildlife-rich days of the past.

In 1935, Aldo Leopold sent Hammy a letter from G. W. Taft, the Necedah postmaster. Taft hoped that the area could be included in the federal plan: "I am . . . much interested in . . . propagation of migratory and upland wild bird life. . . . [The land] has been included into several drainage districts and due to this have [*sic*] been made into deserts. . . . Our water levels have been lowered at least ten to twenty feet and by so doing this vast natural sponge has been completely dried up. It is not so many years ago that we used to have a wonderful flight of wild water fowl, but now along with the blueberry are a thing of the past."[8]

Soon dust storms made that unpromising situation desperate. Goc's vivid re-creation of those days is an engrossing reliving of the notorious May storm that had struck Kansas, Texas, and Oklahoma, and moved northward into central Wisconsin, whose stretches of marsh and oak savannas had previously limited wind damage. A dry winter caused the snow to freeze deep; cover crops of rye were crippled; spring moisture ran off rather than sinking in. Drought followed April's rains—a drought that created exactly the bare, parched land whose loose grains were sure to move and blow in that terrible wind.

"The dark storm hid the smoke of the fire that consumed fifteen thousand acres and twenty farmsteads situated on a great drained marsh in Jackson County.[9] The winds blew for two days and a night, adding to the thick cloud that carried three hundred million tons of dirt all the way to the Atlantic seaboard. Dust bowl folklore maintains that President Franklin Delano Roosevelt felt Wisconsin dust on his White House carpet."[10]

Storms continued to plague the area. Sand buried fence rows and drifted high around the trunks of trees, smothering them. Fran would later watch housewives literally shoveling out their kitchens and back stoops, and snowplows clearing man-high drifts off roads in the aptly named hamlet of Big Flats in Adams County. The Juneau County agent recorded the thirty thousand acres of field crops destroyed by the May sandstorm. By the time it had passed, a distressing amount of Wisconsin sand and precious, limited organic matter and nutrients vanished as nearly a foot of earth disappeared from twenty thousand acres, and almost two million acres became the largest parcel of badly wind-eroded land east of the Mississippi. Experts declared the area "unfit for cultivation."

In Juneau County, state and federal purchases bought most of the four fortunate townships deemed suitable for wetland restoration. The unfortunate township of Armenia, as sandy and damaged as its neighbor to the north, was not included. Without enough wetland to make restoration worthwhile, it remained so desolate that in 1947 it became a bombing range for the Wisconsin National Guard. In Necedah, one hundred families—of whom only four would remain—lived on 4,800 acres. Resettlement would cost $3,200 per family to remove those remnants of the hopeful poor who farmed there, thinking that they could grub out a living. When low fertility, fire, and frequent unexpected frosts made good crops scarce, the farmers resorted to cutting railroad ties or harvesting the bounty of the marshland—sphagnum moss, blueberries, marsh hay, and wiregrass, used in rug making, until the market disappeared.[11] Wives canned what they grew and gathered young fiddlehead ferns in the spring. Children grew heartily sick of picking blueberries, blackberries, and what they called blackcaps—the small, seedy but tasty wild black raspberries that loaded the dense thickets of cane in good years. Men went "shining" or "jacklighting" deer at night, making sure that their shots counted and the illegal venison made it to their tables. These tactics enabled some to hang on.[12]

These farmers were sturdy and self-reliant people—once city people, for the most part. They had come from "from large industrial centers," as Hammy noted, and had started as greenhorns, knowing little of fertilizer, liming, and cover crops. They quickly learned to manage. Life was rudimentary. Only one in ten of their widely spaced homesteads were on improved roads, whereas elsewhere in the state, half of the farms fronted on graveled, graded roadways. Their houses, hastily built of rough sawn white pine and plastered with coarse sand plaster, were kept bearable through the bitter winters and late springs with endless cords of oak, which were cut and split and stacked for the woodstoves. They were served by outhouses and dooryard pumps that often froze in winter. They found what medical and social services they could in small, insular hamlets. Their children attended weather-beaten one-room schoolhouses; a meager trickle of eighth-graders went to distant high schools if they were able to arrange to board "in the village."

Everyone worked. Long winter evenings gave the women time to quilt, mend, and make over worn-out clothing. Families gathered around the stove to crack black walnuts and hickory nuts. Bruised fruit became fruit butter. Children worked alongside of their parents, and nobody complained when school was dismissed for deer season or the potato harvest. Fortunately, "taters"—the staple food—kept all winter: a family might store up to eight hundred pounds on the earth floor of the cellar. As a village plumber observed, "They ate anything that moved—coon, turtle, squirrel, frogs—even cranes."

Toughened by adversity, scornful of softness, and intolerant of airs, these "sand landers" possessed a perhaps extreme form of Depression-era virtues: a powerful work ethic, frugality, and the ability to "make do." Pride and independence were etched deep into local consciousness. To "go on the town" (accept welfare) was a disgrace. Reading a letter preserved by historians of the Resettlement Administration delineates in sharp, painful detail the plight, pride, and worth of one beset by bad luck and Depression economics.

I worked more or less on the home place, till I was 28 years old, then went into the spray game. Made good until the depression came. Everything went on the books, more or less, mostly more. Can't expect to collect it any more. . . . [I] made work for myself by buying the acres . . . not ideal farm, a lot of wood on it . . . Wood dropped to $3.25 the same fall . . . couldn't get help so was out of luck . . . a federal man came around . . . by having our place recommended for zoning by the town board . . . we [could be] moved the following summer. . . . Advised not to clear any more land but keep on farming what we had. . . . Planted spuds. Bottom went out of market. I lost $100. . . . Was unable to pay seed loan. Took list to chairman of groceries needed. Told me to take it to the relief. I asked him why I should go on relief. I saw a lawyer and found I could do nothing. Was compelled to go on relief and was ashamed of it. . . . Had to sell cows, got an average of about $20 apiece for them. I would not take $100 for them now, high grade, high testing cattle. . . . A year ago last winter with a family, 9 months old baby, another child on the way, I got sick for a week. No road. Heavy snow up to hips. Wife couldn't go. I got out of a sick bed and walked 4 miles through that deep snow to borrow groceries to get along. I had to sit down in snow several times and rest on the way back. Just was able to make it—cold freezing weather. . . . If I should not of made it, what would of happened to my wife and child out there alone where no one showed up for a week or two at a time. Please . . . consider my case. . . . I am not on relief and have not been since last fall, and do not want any more of it, if there is any possible way out.[13]

The many meetings held in the area in 1934 by University of Wisconsin Agricultural Extension staff may well have led town governments to encourage the planned departures, but it was this man's desperate circumstances, not plans, promises, or rhetoric, that brought him to seek "any possible way out."

The human conditions of hardships and despair, so evident in the situations the Hamerstroms saw, influenced them. Fran felt the same reaction to economic disaster as she had about the catastrophic effect that the Martha's Vineyard fire had had on the heath hen. She was indignant. The pride of class displayed in her reactions to dark-skinned hunters in Clinton dissolved in the face of this desperate poverty and misfortune. Hammy admired the skills and resourcefulness of the people with whom he worked. The wretched environmental situation motivated these initiates in the science of game management to use field research to improve conditions. Their most significant learning, however, was of the frailty and error that human effort brought to problems on the scale of the ruined township of Necedah and of the deficiencies in their readiness for such a project.

In the late fall of 1935 the Hamerstroms simply transplanted their small store of belongings to a conveniently empty old farmhouse with a woodstove, a sound roof and windows, and a handy shed in the dooryard for the load of cut oak they bought from a neighbor. They unpacked and arranged their books and their field gear. Fran lingered a bit as she stored their city clothes in boxes in the attic; she placed one white box, tied with a ribbon, on top of the stack. It contained a carefully folded burgundy evening gown.

Hammy was Project Game Manager at a salary of $2,400 a year. His colleague, Fred Zimmerman, recollected with a smile, "We called him Fritz. They were one happy couple: he had a degree, and a job. . . . I came in one day when only her Mother was there. She had been crying—I think on account of the house."[14]

The farmhouse near Necedah that was the 1935–1937 Hamerstrom home, during the Resettlement Administration.

Mrs. Flint could hardly rejoice about the bleak and daunting situation the young couple faced. The drought had caused conditions severe enough to attract national reporters. One famous journalist wrote of Wisconsin's "pinch-faced children," and "starving, bawling cattle."[15] Even with a helpful rain in June after the great May storm, three-quarters of the farmers in neighboring Adams county and half of those in Portage County qualified for drought relief. Modest payments from counties encouraged soil conservation practices; the hourly pay offered for planting shelterbelts with state-supplied seedlings was a godsend to the many farmers who could not qualify for the payments for the required practices. Extraordinary efforts by county agents— even when the lack of county revenues to pay them threatened their jobs—improved the lives of some residents. They could, at least, increase the yields of home gardens and organize entertainments to lighten the dismal times. But the underlying problems remained. As Floyd Reid, sturdy survivor of those times,[16] testified, "We were all poor, but some were dirt poor. They just couldn't afford what the county agents said to do."

Fran, in her distinctive way, remarked tartly to me when she read a draft of this chapter, "You see those men as heroic; we saw them as villains! They helped destroy many of the wild places we wanted to save." That reaction reflected years of seeing woods cleared and marshes drained in the land surrounding their Plainfield home as well as her personal response to the times. But it also illustrates exactly the gulf between the expectations of the ahead-of-their-time environmentalists and those of local people. That gulf continued throughout their lives and indeed, continues today. Recalling the Great Depression gives evidence of intractable discord and many disappointments. Often thwarted as director of the Resettlement Administration, Rexford Tugwell replied to a friend's criticism of his effort as pretty conservative by saying, "Beanie, we were pikers."[17] And the extremes of feeling about Franklin Roosevelt and his programs—from hatred to near idolatry—have scarcely been matched in recent times.[18]

In the dismal and divided year of 1935, Hammy set to work with characteristic thoroughness and zeal. First he needed to establish what wildlife was still present. Not only groups of species, such as avian predators and furbearers, but individual ones—prairie chickens, sharp-tail and ruffed grouse, sandhill cranes, horned owls, white-tailed deer —had to be surveyed. Recording their findings would require maps, and since accurate maps of the site did not exist at that time they were required walk or explore for many miles. "Any trail you can travel by car is a road," said W. T. Cox, former Minnesota state forester, who was in charge of forestry and wildlife for the whole site.[19] Fran helped make the first real road map of part of the area, a vast expanse of interlocking trails, which often seemed to lead nowhere in that thinly populated flatness. The Hamerstroms and their crews mapped 7,400 acres.

Hammy wrote for advice to Franklin S. Henika, the Madison-based Wisconsin Conservation Department's regional game manager. He had completed mapping the main roads. Must he include the hundreds of hay trails? It would take enormous amounts of time, even should his men be able to survey some species and cover mapping at the same time. He had collected one hundred plants along with a variety of seeds and intended to follow up on Leopold's early investigations into prairie chicken behavior.[20]

He established liaison with the Black River Falls site; certain tasks applied to this site as well as the Necedah location. He found good colleagues there, such as Fred Zimmerman, who, like the Hamerstroms, went on to make a career in conservation. Zimmerman speaks proudly of his service in the Resettlement Administration: "We did worthwhile work. We picked places for future state parks, set them aside as conservation areas and improved them. There was lots of tree planting, mostly of jack and Norway pine. We got work crews from the Mission to the Winnebago. Some settlers were successfully moved to Trempeleau and Dane counties. We'd get together at meetings, hear reports and share ideas. . . . [We had] a congenial team."

Zimmerman, who saved enough from his salary of $166.66 a month to finance his Ph.D. studies, was keenly aware of the contrast between himself and the families around him who lived on $150 a year. Fran and Hammy joined that isolated farm population. Their new neighbors walked through the woods carrying presents for the newcomers: winter squash, pumpkins, potatoes, and citron, which Hammy had to identify for Fran—she knew only the product that Cook had chopped for fruitcakes. They asked the childless newcomers for advice in unexpected areas. How had they avoided having children? There were far too many children; many of them died.

"We loved those people," Fran insisted. "It wasn't long before Hammy was fighting for them. Sitting at our kitchen table, he would compose the letters they asked him to write. Some of the Poles were uncertain of their English, others, years away from grammar school, had replaced book learning with survival skills."

Hammy realized that bureaucratic detail would overwhelm him unless he designed an effort for his studies. A chance meeting showed him a way out. On New Year's Day 1936, he chatted with the truck driver for a crew of laborers out of Black River Falls, a man named Os Mattson. Hammy liked his straightforward ways and liked even more his knowledge about game. Fran needed a helper; Hammy assigned him to her crew. She, as an unpaid worker, was collecting and analyzing owl pellets. She noted his keen observations. "He learns quickly," she reported, "and makes good suggestions." Hammy made him foreman, which meant a salary of $50 a month instead of $40.

Soon Hammy had three "foremen": Os, Bud Truax, and Jimmy Blake, chosen ostensibly to direct others but really to provide him with a research crew. In effect, he and Fran were running an informal school of wildlife research methodology. He and his counterpart in Black River Falls set up a "research and investigational division" ("if we may be called such," as his report wryly stated) in April 1936, using both Truax and Blake, plus Zimmerman, and two biologists from the Necedah site, half time.[21]

The new team did an astonishing amount of work. They organized trapping and banding programs; established experimental food patches, fallow plots, and cover plantings; and planned for the winter-feeding for a variety of wildlife. They painstakingly evaluated, tabulated, and compared cultivated corn, millet, buckwheat, and soybeans to food-weeds grown on thirty-five experimental plots, some plowed, some plowed and disked, and some left fallow. They located nest sites of pheasants and grouse over 24,900 acres. They found eighty-six booming grounds, where prairie chicken cocks conducted their spring displays; they plotted the areas most favored by the birds of prey—owls and hawks of every variety—and searched the ditches for otters, muskrat, mink, fox, and other furbearers. They mapped and color-coded every sighting of each sharp-tail and ruffed grouse seen in two years. Hammy collected plants and located plant communities.

Hammy distilled and organized all this information into a final management plan of twenty pages of densely packed overview and tables, followed by twenty-three separate detailed reports on the surveys and programs he had initiated. The implications, he wrote, might well be national in scope.

He did not mention the fundamental problem in the Resettlement design, although it was clear to his discerning eye. Cox had asked Hammy to pay particular attention to fire lanes, cut in an attempt to control the ever-threatening fires. "They crisscrossed the area with them, and some of them are useless. Tell me what you find." Hammy found certain lanes impassible. He walked the "most uninviting" soft sand ones or inspected them through field glasses. He found no

Fran, "our ex officio member" as Hammy called her, at work with foreman Os
Mattson, assigned by Hammy to be his wife's aide. Os later became their right-
hand man in the early years at Plainfield.

warning signs for missing bridges, no consideration of existing ditches
and proposed flooding. He questioned: surely missing bridges were
hazards for firefighters? Who would maintain the lanes after the relief
labor camps were abandoned—as they were, in 1937? [22]

"Relief labor camps." The phrase speaks directly to the basic
dilemma of the Resettlement Administration. Even the workers saw
the flaw at once. "It was a welfare project," declared forthright Os.
"The idea was to keep the men occupied. We were doing things that
didn't matter a damn. We built quail shelters. The quail didn't use
them. And big deer shelters out of jack pine. The deer never came near
them." Fred Zimmerman agreed. His careful records of two winters of
feeding proved that shelters shaped roughly like wagon wheels were
of "no use whatever" but that racks—crude adaptations of farmer's
racks for feeding hay—brush piles, windfall shelters, and lean-tos
could be effective if well located. No one knew exactly what would

work; high rates of disappointment and failure were the rule for those early practices of conservation research.

Learning was synergistic: the foremen passed on improved methods to the various crews. Hammy learned from these country people, whose reliance on game for food taught them much. Fran especially lauded Jimmy Blake. "He was half-Indian, handsome, erect, and tall—and he washed his clothes in the spring. How much we learned about tracking, moving, and finding animals from him!" He was a far better woodsman than either of them were. Hammy soon put Blake's knowledge to work, putting him in charge of three crews of two men each.[23]

Blake lived in Necedah. His considerable abilities were lost to the project when his wife insisted that he move to Chicago to make money in a factory job there. Fran grieved. "He knew all about moving quietly through the woods and what signs to watch for." Hunting with him gave her new insights. Os Mattson recalled an argument about shooting ducks on the water.

"I'd never do that," exclaimed Fran, shocked at the violation of sporting practice. "If you was counting shells," said Jimmy, "you wouldn't be practicing all them wing shots." That lesson in human understanding stuck. The sporting rules of aristocrats could not be imposed on poverty-stricken sand landers.

Hammy's meticulous instructions for the fur survey maximized the value of his team's keen observational powers. Pairs of men, supplied with the essential compass and a stenographer's notebook, walked designated ditches, streams, flowages, and marshes. On the top page of the open notebook, inscribed with the date and location (town, range, and section), they were to sketch landmarks such as roads or notable trees, at a scale of one inch per mile. On the page below they wrote field notes describing the fur signs found: the traces, dens, droppings, tracks, tufts of fur, remnants of meals and the like left by skunks, otter, fox, beaver, and "rats," as muskrat were commonly called, or even, in those days, wolves and badger. The result—Jimmy Blake's notes—still make good reading, just as they were written.

May 6. Sec.3. T19N-R2E. Saw 3 deer along ditch. Some old beaver signs and 1 small beaver dam. It seems that at this time . . . rats and mink are to be found mostly along small and stub ditches . . . [where] there is better feed for both, as the rats get young grass sprouts and water bulbs that do not grow in deep water. The mink feed on the frogs and minnows that are plentiful in the small streams. Deer are changing from a dark gray to a dark yellowish color. We can nearly step on grouse before they fly but can't locate a nest.

June 2. Sec. 23, T19N-RIE. Skunks dug out a nest of turtle eggs on ditch bank, and ate them. Saw one sharp-tail and one yellow leg. June 4. Sec 33. T21N R1E. Four beaver houses—one extra large. 3 beaver dens built in ditch bank. 4 beaver dens and very much cuttings along ditch. One otter slide and droppings. One large rat house. Saw 8 sharp-tail along fir lane west of Bear Bluff. Also one fresh wolf track and much wolf sign, along the fire lane and around the bluff.

June 14. Sec. 2. T20N-R1E. Caught and examined a spotted fawn and let it go. It stood about 1 1/2 feet high and was about 22 inches long. Legs were about the size of a man's small finger and hoofs . . . the size of a man's small fingernail. Weight–about 8 pounds. Was very small and staggered when it ran, got caught in a bunch of small willows and could hardly get free. Was lying down in some tall grass, but out where the sun could beat down upon it. Got within three feet of it and it never moved except to get its head as tight to the ground as possible.

June 18. Sec. 1. T19N-R2E. No change on Little Yellow ditch. Fresh badger and raccoon tracks along ditch, no other change. Saw a marsh hawk swoop down and light on a small knoll. When I got over to it I found it had destroyed the nest of a bird, killing four young ones. It had partly eaten three of the birds. The birds were small, about the size of a young robin. Had no feathers except a fringe along the back of the wings, a light blue color. Nest was made of fine grass and was well concealed in some rather tall, heavy grass, on the ground. Rat and skunk sign on ditch going through southwest corner.

It was in this natural way that Hammy learned the talents of these ordinary people, their comfort with the wild creatures that had always peopled their environment, and their habit of paying attention to them. Their font of homey knowledge informed his research, and the resulting painstaking record made clear the parlous condition in which he found the area wildlife. "The game and fur populations . . . are sadly in need of assistance," he warned. Encroaching brush gave fox and sharp-tail grouse the advantage over prairie chickens; low water explained the diminished numbers of waterfowl and muskrat. Deer and quail were increasing, deer to the point of winter starvation, but populations subject to cycles (rabbits, ruffed grouse, and prairie chickens) needed further study. Otters, mink, skunk, badger, foxes, and coyotes were noted.[24] It seemed urgent to move ahead, with energy and dispatch.

That first year was one of discovery and excitement; their home, however primitive, was harmonious and always lively. Guests never forgot Fran's 1936 Thanksgiving dinner. Holidays meant little change in her routine: she still had owl pellets to go through. Nevertheless, she invited the other biologists to a celebration. They arrived to find her sorting the pellets on a small table covered with little piles of fur, bone shards, and the crisp remnants of insects. She alternated this task with kitchen chores, continuing her separating and sorting, and then, as one guest later said in humorous resignation, "She would go back to her cooking." A permanent part of the Hamerstrom legend rose from such details, remembered, sometimes with shock or distress but often fondly, by those with whom they came in contact.

Visitors, including Paul Errington, came to their ramshackle house. His scientific caution puzzled Davis, visiting once with Fran's younger brothers. "I noticed he never said a simple yes or no. It was always 'I strongly suspect.' That drove me crazy."

Fran was content. In those days of anti-nepotism rules and a bias against married women taking a man's job,[25] she did research with her own crew. She embarked on a study of raptors. She found marsh hawks (harriers, as they are now called) in residence, counted migrating redtails, and noted the less common roughlegs, goshawks, and smaller species: kestrels, sharpshins, and Cooper's hawks.

Again, Fran was determined to prove herself—to Hammy's colleagues, to the public, and to the men, and no task was too daunting. Regulations mandated working in pairs. One winter morning Hammy sent her on a game survey alone. He gave her a map showing a blazed trail going across a stream; she would not show him the apprehension she felt. When she came to the stream she saw she had to give up or swim. She stripped, swam, dressed, and then ran to warm up. Avid to master all possible ways of educating the public, she began contributing brief educational notes to the local newspaper concerning various conservation topics such as the deer glut and the ethics of hunting. Her reward was her own satisfaction, and Hammy's praise. "The services of our ex officio member, Mrs. Hamerstrom, have been of such value that it would be most improper not to acknowledge them," her husband wrote in his formal report to administrators.

Both of them reveled in their work and in the daily riches of their environs—the new sprouts of grass, the signs of playful otters, the muskrats carrying grass in their mouths, the occasional wolf tracks, and the glassy water behind beaver dams. Pleasures such as these brought deep satisfaction. Hammy wrote to his parents: "This country . . . called barren by many . . . is beautiful. Mixed jack pine and scrub oak, with some other pines and oaks, on the sand islands, broken up by innumerable grassy meadows and marshes; in the creek bottoms bunchberry, Clintonia, waist-high ferns, virginia creeper, birches, maples, alders, willows. Everywhere purple and white asters, goldenrod, and blazing star, many blueberry patches, some wild cranberry, leatherleaf bogs and an infinite variety of plants new to us. . . . There are sphagnum beds and small remnants of tamarack swamps. Even the solid stands of popple—which come in after fires—are simply lovely in the purple light of winter evening."[26]

That delight in nature and love of the land sustained and energized them, even as they became increasingly aware of the failings of the Resettlement Administration. Fran was furious when, "they sent photographers to record their accomplishments so they asked us where they could find the worst shanties, the poorest families. One mother complained to me, 'Those men wouldn't even let me wash the childrens' jaws!' They whizzed right by the [few] more prosperous farm houses."

Since the sand county farmers could only be moved if they chose to go, local bureaucrats with federal goals to meet made promises they could not always keep. Farmers paid the price. One family on the Hoffman marsh agreed to sell. Expecting a quick move, they sold their small store of machinery and their livestock, down to the chickens—then waited for payment through a long, harsh winter without cash or sufficient food. Others waited over a year for the settlement.

Chris Mattson was one who refused to sell. "I love this place. I can do with $150 cash a year. I'm gonna make it." Then he learned his wife had cancer. Grimly he marched into the office in Necedah and told the official he would sign the papers. "But I need the money right away.

I gotta get my wife to Mayo for treatment." The official reassured him, it would not take long. But months went by, the money did not come; the trip to the clinic could not be made in time. "That beautiful young woman died," recalled Fran, still unresigned to bureaucratic bungling over fifty years later.

The Hamerstroms determined to right such wrongs. In 1936 they wrote a scathing expose. They took it to Aldo Leopold, with whom they often consulted.

"Who is this for?" he inquired after reading it.

"The *Saturday Evening Post.*"

"You are right, of course. This is well written." He paused. "It would surely have an impact." He refilled his pipe and tamped it down deliberately before speaking.

"But I'm going to ask you not to publish it."

Leopold was telling them that they were too new, too unknown to become crusaders. "Leopold," said Fran later, "had an uncanny sense of timing." Silently, doggedly, they returned to the farmhouse to the simple life they had sought.

Through the harsh winter, like their neighbors, they cut wood with a two-man saw, pumped water into zinc buckets, closed off all the rooms except the kitchen to conserve heat, and adjusted to the isolation and work load. When snowbound they did the game census on snowshoes. The routine was unvarying and dirt-simple: load wood, stoke the fire, rack your brain about what to cook with the remaining staples, put on layers of clothes, socks, scarves, hats, gloves; bind on snowshoes; come home rosy-cheeked and exhilarated, write up your field notes, warm up yesterday's stew, fall asleep over supper, stoke the fire, crawl into bed. As they observed, "There were good books in the house for the long winter evenings. They stayed on the shelf unread."[27]

One winter the pump froze solid. Nothing they tried—even using up too much of their precious stock of wood to melt snow—made the slightest difference. A neighbor boy who happened by asked for rags and kerosene. Debutantes didn't save rags, so Fran had to rummage through boxes in the attic to retrieve something they could spare. Slowly she unwrapped the wine-red velvet evening dress that had always been Hammy's favorite.

I would never wear it again. I hurried downstairs whistling and handed it to Andrew. . . . He nodded and wrapped my dress around the pump . . . poured kerosene, soaking the dress again and again. Then he reached in his shirt pocket for a match. I went back into the kitchen and watched the flames leap high. . . . Frederick stood a little apart, aristocratic and elegant, even in his worn field clothes. I could see his fine black eyebrows and his small straight nose. He wasn't watching the flames. He watched my dress, and when it was gone, he still looked at the pump. His face was inscrutable.

That blaze at the pump marked a Hamerstrom watershed. Their life was here; they would never go back to that "rather charming mode of life we had known before . . . that odd little half hope that we might combine the old with the new."[28] Instead, somewhere, they would manage the public domain so that sandhill cranes and chickens would be plentiful, the fur harvest planned, and the deer population controlled. It would take many years of further experience to teach them that success would come in much more focused, more species-specific, endeavors.

Hammy knew, perhaps as early as the first winter, that Necedah was not likely to be the place for him. The rules and strictures of distant bureaucrats limited him. When, for example, delivery of tools for the workers he supervised was delayed for months he could do nothing except impose what was called a "continuous game inventory," which required the work crews to tramp in long lines through the woods, more like game beaters than surveyors, counting whatever game they came across. Of course, in noisy formation, they saw little or none.

The frustrations mounted. Perhaps the last straw came in January 1937, in the very depth of the Great Depression. Hammy had painstakingly listed trees and shrubs for spring planting to supplement existing stock of wild grape, woodbine, and prickly ash. He asked for 500 thornapples, 250 black locust, 750 junipers, 750 sumac, 750 wild plum, 4,500 maples, and 5,000 white cedars to "serve as nuclei for natural dispersal . . . [and] spreading" of species that would provide feed and shelter for wildlife.[29] The regional office delayed Necedah's order; his stock arrived long past the planting date and shriveled in the summer heat. Then the vital, ongoing mapping work slowed. Accurate mapping depends on trained use of the compass, and the indispensable compass men[30] were not on the relief rolls. Men on relief had hiring priority ("work, not welfare"), so the experienced mappers had to be let go.

Throughout these years they attended Aldo Leopold's seminars in Madison. In the snowy month of March 1936, F. S. Henika, Wallace Grange, Forester Cox, and W. F. Grimmer went with Hammy to report on their investigations. "Great speakers!" said Leopold. In April he described their trapping, banding and feather-marking plans; and later, with fellow graduate student Art Hawkins, he reported on a canary grass project.[31]

To the Hamerstroms' relief—indeed, perhaps at their request— Leopold paid the project an official visit. (It may have been about that time that he encouraged Hammy to join his fledging graduate program in his new game management department in Madison.) Certainly his blunt report, following his visit to Black River Falls validated Hammy's dissatisfaction. It was a pity, Leopold wrote, that after an expensive deer census, no kill record was kept in 1936. "Plain waste," he said. "It is a loophole you could throw a cat through." The Resettlement Program was too complex and many-sided for one man to be aware of another's findings. Leopold called for a policy that would support, indeed require, publication of research findings. As he said, "One effect of not insisting on publication is that you have gathered a monumental mass of data, little of which is shaken down into usable theories, policies, or rules of practice. This always happens when fact gathering is not accompanied by digestion of the data. Publication . . . [forces] digestion."[32]

Leopold's recommendation was ignored. Hammy's detailed studies were judged "unsuitable" and too expensive. A pragmatic supervisor, the man who had to field public complaints, suggested that such investigations belonged at a university, not in an agency designed for a relief program. The cost of good scientific studies could provide ammunition for critics of the resettlement effort. Practical improvements would win support.

The Hamerstroms were to hear this call for useful work over and over again in the years to come. The arguments were always the same. Improved trout streams would meet public approval, but research that was abstract or weighted toward the theoretical was received with suspicion. Visible achievements were the way to get the support of the ordinary, everyday taxpayers. A disheartened Hammy wrote his parents: "I grow daily more discouraged with the mess which the government insists upon making of this job, and see the chance of any worthwhile result becoming more and more remote. We have made arrangements to go to the University to work under Leopold next autumn, and would pull out of this present mix-up tomorrow if we felt we could do so with dignity. . . . It has been made very plain that this is primarily a relief proposition which means that any benefits to game are of rather secondary importance . . . about the most useful thing I can accomplish is to prevent as many of the most blatant mistakes as possible—and there is little enough I can do there. . . . At least we are saving a little money toward our graduate work."[33]

Years later Hammy spoke his mind about the Resettlement Administration. Its governing principle was totally unscientific. A senator in Washington might call the headquarters with the request to return Necedah to the pristine waterfowl area it once had been. The district and regional offices would promptly mandate a priority for creating waterfowl habitat. "It was perfectly plain by then that prairie chickens were having trouble and were losing range very rapidly, but flooding out the best prairie chicken meadows to make waterfowl ponds was quickly approved." A subdued shudder accompanied his final word: "Dreadful!"

Clearly, in 1937 neither the politician nor the public was ready for research-managed wildlife projects. Small wonder that in a poverty-stricken backwater of Wisconsin, few understood that habitat (a word not then in common use) could be maintained and restored only after scrupulous study of the flora, fauna, and human context of an area.

Hammy and Fran knew now how much more they needed to know. Their observations of prairie chickens, for example, generated many questions. Where did the birds hide their broods? Where and how far did they go in the winter? What could be learned from the booming, that resounding spring ritual that governed the mating and propagation of the species? During their ensuing three years of study at Madison, they answered some of those question and generated many more.

They would take satisfaction in the establishment of the Necedah Migratory Waterfowl Refuge in 1939, soon to become The Necedah

National Refuge, where Hammy's goals were at least partially accomplished over many years. By the 1960s the site of Hammy's early endeavors covered several townships. Today, some forty-three thousand acres in Wood and Juneau counties are managed with substantial and varied provision for wildlife by county, state, and federal agencies. The refuge hosts well over a hundred thousand visitors a year—mostly birdwatchers and nature lovers, although regulated hunting, fishing in the summer and winter, trapping, and berry picking are permitted in season.[34]

Over time, then, some of Hammy's priorities have been reflected. That the Hamerstroms are not mentioned in current public information on the refuge is understandable; it was not designated as one of more than five hundred national refuges[35] until well after their departure. Still, people living in the area today remember them. Some follow, at least to some degree, their example.

Nevertheless, the years from 1935 to 1937 were invaluable for the two of them. Working with their hands and brains in the wild place they loved, they lived the necessities of disseminating the results of thorough research. The human tragedies they saw deepened their natural empathy with the underdog and left them with a permanent distrust of large bureaucracies. The vulnerability of superiors taught them the limits of possible action and the crucial need for conservation education. The forced intimacy of isolation had served to increase their belief in themselves and trust in each other. Fran had become a fellow researcher, with her own favorite species to study and a crew of his best men to supervise. She had proved her worth and had begun to write, however humbly, for the public. All of that would serve them well.

They had both been accepted into Leopold's graduate programs well ahead of the day that they packed their belongings into the familiar trailer. That warm early fall day in 1937 they said goodbye to their friends and neighbors, hitched the trailer to the Essex, and headed down the country roads, across the Merrimac on the small ferry, down to Madison.

# 6

# Enter Leopold and the Chickens

*A teacher affects eternity; he can never tell where his influence stops.*
—Henry Adams, *The Education of Henry Adams*

Aldo Leopold, whose profound effect on the practices of conservation and game management continues to this day, believed in a "new social concept toward which conservation is groping."[1] Ethics, he declared, must guide our dealings with nature—an idea that was new to most Americans of the time and accepted by even fewer. To most citizens, conservation meant damage control, then beginning to be provided by the 840,000 acres in twenty-two new game refuges across the nation.[2]

Leopold, a graduate of Yale forest school, worked first for the forest service in New Mexico. There he began to move ahead of the attitudes of his time. In the 1920s he saw planting crews ruin a rare open patch of clover—ideal food for deer and partridge—by planting seedling pines over it and became increasingly aware of the importance of predation to the health of the environment.[3] After a brief assignment to the Forest Products Laboratory in Madison, Wisconsin, he undertook a survey of the north central states in 1930 for the Sporting Arms and Ammunition Manufacturers' Institute. He recognized the need for a new breed of managers who would devote themselves to research-based game management. To educate them he proposed that the Wisconsin Alumni Research Foundation (WARF) support a new course of study at the University of Wisconsin. The foundation agreed and in 1933 funded a five-year budget of $8,000 a year, which paid Leopold's salary and modest program expenses.

Leopold was seen as a somewhat unorthodox academic in an

unproven profession. Although he "lacked the requisite Ph.D.,"[4] he was appointed by the foresighted Harry L. Russell, dean of the College of Agriculture, to head the new entity and serve as its sole faculty member. Formal designation of a department had to wait until 1939. In the early years, Leopold was housed in an office in the basement of the Soils Building.

Each year Leopold accepted only five students who "could and would think."[5] He wanted independent people with considerable field experience who would not only be excited about doing research but remain open to new ideas. Both Hamerstroms met the specifications that seemed to have been tailor-made for them. They joined the program in 1937, Hammy as a Ph.D. candidate and Fran, to her husband's deep satisfaction, for the master's degree.

They chose to live in an old farmhouse on a ninety-acre farm close to Madison. It needed repairs but was rent free; they settled in. There in that gently rolling Dane County countryside they missed the tranquil moonlit nights of Necedah, where the sound of neighbor Jack Becker's baritone had floated into their bedroom. "Do you miss our neighbors, Hammy?" she asked him.

"I do," he replied. "But I don't miss the tractor accompaniment to his singing while the poor devil plowed half the night after working in town all day!" But in fact, they reveled in the stimulation of a university community and the congeniality of their new situation. Chicago offered visits with the Darrows and meetings of the American Ornithological Union. They went dancing with their friends and made new acquaintances, one of whom was Roger Tory Petersen.[6]

Leopold soon gave Hammy the title "Assistant in Game Management" and charged him with handling correspondence for the cooperative bird-banding group he had created. This involved letters to and from men in Wisconsin, Michigan, Ohio State, North Dakota, Minnesota, and Illinois. Results were not remarkable—Hammy received many answers to his requests for reports that said no banding had been done, or that birds did not respond to bait in mild winters, or that funds had not been provided—but his experience and the resulting network gave him a foundation for future work.

On a trip to New Orleans in the early 1940s, the Hamerstroms met the director of research for the Louisiana Conservation Department.

He entertained them in "the oldest apartment building in America" on the Place d'Armes, and he took them to the delta at the mouth of the Mississippi River.

Hammy described that experience in a jubilant letter to his parents. At that time, several of the then unpopulated arms of the delta lay within seventy miles of the New Orleans. They explored the branching channels, the silt and sand drops, the higher ground near the river, and the huge marshes with small ponds and meandering canals. "And the craziest part of the whole crazy business is that one can follow side channels clear to the Gulf in fairly deep water . . . as the silt ladened water hits the open Gulf it spreads out, slows down, and starts to deposit. It's quite a trick to keep the navigation channels deep enough, and piloting is a very fine art. . . . We saw geese by the tens of thousands. I would'nt have cared if I had'nt fired a shot. As it was, we had one day of marvelous hunting . . . . I'll never forget it."

In Madison, they experienced Leopold's Socratic teaching methods. He demanded curiosity and critical thought applied to facts and the ability to relate them to theories and experiments in land use. "You spoke from the data, and if you didn't have enough data, then you were wide, wide open, and you *got* it."[7] Leopold did not discourage acting on hunches, but he insisted on careful observations. "Read the landscape!"[8] Hammy found good use for his expertise in identifying the native fauna and flora.

Leopold's biweekly noncredit seminars were a ritual. He set a constructive tone for student reports on fieldwork assignments. "There was that big, delighted smile when you returned from the field." He passed around beer, or apples. Some students told of projects at the Fabrice farm near Whitewater or the Faville Grove Wildlife Center where they worked cooperatively with farmers to improve habitat and encourage wildlife. Fran found the rivalry "fierce"; others called it subtle.

Leopold wanted to raise public acceptance of conservation as the way to a better life for all. Understanding that "The reaction of land to occupancy determines the nature and duration of civilization," might take "generations rather than years,"[9] he felt sure that habitat improvement would come as facts were seen and acted upon. Healthy habitat would include predators and make use of edges: those humble borders of farm fields—fencerows, woodlots, and wetlands—that sheltered and nurtured wildlife. He recognized that the increasing size of farms, and the growing practice of clean farming, would threaten those important keys to game productivity. He expected that flexibility would be required as farming changed. (Leopold would, I surmise, have strongly approved the current recommendation to leave the corners of fields irrigated by large pivot systems unplanted to row crops, so that wildlife can find shelter there.)

He broadcast his message in varied ways, regularly teaching a January short course for young farmers and working with horticulturists and botanists at the university arboretum. He gave short, meaty talks over WHA radio, the wide-reaching university station, and presented appropriately crafted speeches to groups of all kinds and sizes.[10] Neither Hammy nor Fran ever forgot his powerful example. Fran, with an apparent inborn flair for publicity, was more likely to explore new and novel ways to educate than was her more conventional husband.

Leopold, fondly known as "the professor," trained his students by assigning to them all arrangements for the 1937 Midwest Wildlife Conference—an invaluable experience in networking. He advised them not to write out the papers they would present. "Reading kills the material. If you don't know it well enough to speak about it from notes or an outline, you aren't really ready to deal with the ideas."[11]

Students responded to Leopold with respect, affection, and even protectiveness as the often-told story of their finding a regular office for him instead of his basement quarters shows. Tired of making do with limited space in the vestibule they all shared with the campus gardener, the students located an empty facility and moved his office. Fran and Hammy provided the car for that move, and next morning took the astonished professor to his office—completely furnished down to his pipe and flowers, arranged in a mason jar by Fran. A nonplused dean accepted the fait accompli.[12] Not until two weeks later did Leopold's funds for the next year come through.

About writing their mentor was explicit. "You have a long way to go before you can ask anybody to invest his time with this paper," he said to one abashed author, " . . . you can't shift what should be your own hard work on to an editor." His own bulging embossed-leather portfolio, ladened with ripening manuscripts, illustrated the point. "Here's where I put my stuff to simmer. It doesn't change in there, but I do. . . . Think of it this way. In spite of all the . . . advances of modern science, it still takes seven waters to clean spinach for the pot . . . and for all my writing to this day, it still takes seven editings, sometimes seventeen, before I let it go off to press."[13]

Hammy, perfectionist that he was, worked diligently to present his first paper as a polished one. To his chagrin Leopold said, "Frederick, you have a good first draft here." Soon Hammy realized that the professor "had such magic that sitting beside him with the 'perfect' manuscript in front of us, I came to be able to recognize the trouble spots often at the moment he did, even before he spoke."[14] Leopold was to hold up some of Hammy's papers to others as models of thesis work—suitable for publication.[15]

A paper that predicted the population explosion of Wisconsin's deer herd showed Hammy as prescient. It concluded, "There have been no losses from starvation yet. Such losses are more easily prevented than stopped: now is the time to take action."[16] It happened. Deer, having put venison on many a settler's table, moved south into farmlands, causing significant crop losses and numerous automobile accidents. The inevitable starvation in hard winters came; Hammy's reputation for wisdom among hunters and conservationists, who had noted his generally unheeded warning, grew.[17]

A noteworthy aspect of those papers is the identity of the second authors: Millard Truax, on one describing traps, and James Blake, on "A Fur Study Technique," and studies on muskrat and deer. Few would recognize those men as Hammy's former foremen, neither of whom had gone beyond high school. Hammy knew how their woodsman's wisdom had contributed to the papers; they had earned the credit. Hammy always gave recognition where credit was due. Leopold felt the same way; Hammy later quoted his teacher's statement against the demand for accreditation. "Good people" he had declared, "can be trained anywhere. A formal education isn't a prerequisite."[18]

Through the years ahead, Hammy, like Leopold, built from strength. He would always find and rely upon such good people, those whose experience brought essential skills and attitudes to game management and research. This practice troubled some; indeed an academic friend remarked, "He had a remarkable scientific mind—skeptical and critical. Unfortunately, that trait did not carry over to people."[19] Numerous coworkers—and some of today's conservationists—would disagree.[20]

Forthright Hammy also relished Leopold's direct speech. Here was a man who valued integrity as much as he did. He admired Aldo's protest when the president of the powerful National Rifle Association stated that killing eagles in Alaska was "the purest of rifle sports." Aldo reacted: "I would infinitely rather that [the author] shoot the vases off my mantelpiece than the eagles out of my Alaska. I have a part ownership in both. That the Alaska Game Commission elects to put a bounty on the eagle, and not on the vase, has nothing to do with the sportsmanship of either action."[21]

Leopold liked Hammy's natural authority and his academic record and put them to use. Hammy was charged with advising new students. In 1939, Bob McCabe found Hammy seated in the department head's chair.

"Have you had any botany?" inquired Hammy. "It's the prerequisite for the taxonomy of grasses."

"No. I had hoped to continue with biology. I'm particularly interested in wildlife."

"Then you'll need to learn plant species. It's the only way to understand feeding patterns. We'll put you in Botany 1."[22]

In sum, working with Leopold was liberating. "He was a people person. That was his long suit," declared Art Hawkins.[22] His influence often appeared to reinforce a person's innate inclinations. For example, Aldo's frustration with the claims and barriers staked out by academic disciplines may have strengthened Hammy's reluctance to become an academic.[23] He and Fran needed freedom. He would hate the meetings and the politics. It would be stifling to live in a city. Moreover, he saw how inhospitable universities of that day were to women scientists, and he believed that Fran would be happiest in contributing to biological research. "Not for us," he said firmly.

A final affinity between the two men was a mutual interest in prairie chickens; one of the species the Hamerstroms had studied in Necedah. Aldo, who called hunting chickens the "grand opera" of the sport, had always been interested in the birds. He had studied the cycles traced by pioneer researchers A. W. Schorger and Wallace Grange; his own 1929 survey gave him an estimate of 110,000 birds in Wisconsin, almost equally divided between sharp-tails and prairie chickens, and presently he assigned his first graduate student, Franklin Schmidt, to research the state's most concentrated population—in central Wisconsin. Schmidt, and most of his data from a number of years of research, were lost in a farmhouse fire. [24]

The couple was gratified in 1939 when Aldo asked them to continue Schmidt's work.[25] Hammy, with four zoology courses, three in advanced botany, a full year of geography, and a semester of soils had only one remaining requirement, "Agricultural Economics and Wildlife Management." That practicum would give them two years of intensive prairie chickens study, enough, Aldo assured them, to get a splendid thesis from the material. Fran could find some aspect of the project to explore for the required master's thesis. They would be fully engaged.

Leopold outlined the task in his News Letter No. 1 of 4 January 1937. Neither the extent and direction of chicken movements nor the secrets of managing the birds were known, so methods for sexing and banding chickens must be improved. Trapping, banding, feather marking, and releasing the birds must precede any count; hunters' reports on bands throughout the north central states would establish the range. "[If] . . . we can get started and develop the [trapping] technique this winter, then large-scale operations will be possible next year." That was when a cyclical rise in population should occur. The Conservation Department, Soil Conservation Service, and Biological Survey would cooperate, as, he was sure, would farmers and sportsmen.

Leopold would provide prepared feathers for imping, (marking individual birds by gluing a colorful foreign feather into the shaft of one of their tail or pinnae—the feather tufts at the sides of the neck). His graduate students would help with trap design, bands, reporting forms, bibliographies, and publicity, and perhaps someone could provide the plan for the baited tipping platform old-time market hunters used,[26] since Schmidt's design for a "confusion-entrance" trap had vanished with his papers.

"You'll have to start almost from scratch," Leopold warned. "Wallace Grange and Art Hawkins will be helpful." Hawkins had found a few of the elusive creatures on one of Schmidt's research sites. He had ridden the bus for the six-hour trip to Coloma. Farmer Cardo met him and took him out to his farm in a horse-drawn sledge on dirt roads to the location. "They're more than wary," Hawkins warned "and you won't believe how desolate it is up there in the winter."

"That's just what we want," retorted Fran, "a lonely place—in easy reach of colleagues!" Fran was later to summarize her feelings about isolated wild places, Ruthven and Necedah among them. She remembered the elation and stimulus they felt when visitors came and the camaraderie of the occasional work with colleagues, in contrast to being "snow stupid" and dull after days of weather-enforced solitude.

I was very much interested in what you said of Idylls—It is what we have chosen since the beginning (1929 when we met each other). We have always wished to live in lovely wild places. To me, Idyll means a very peaceful place so that is neither quite what we have had nor what we have wanted. It has been a combination of seeing few people—living almost alone—adventure, and above all our work.

We learned to our astonishment that to live all alone is not good! We found we were . . . not being very productive. When we saw people again and they asked us questions we both answered at once (eagerness to talk) and we often used the very same words in answering—just like a chorus. It was very funny.[27]

The constant use of the pronoun "we" announces their shared career aspirations as well as joint pleasure and satisfaction in their activities, however divided. The terms of employment at Necedah had dictated Fran's early subordinate position, as did her less specialized education. Now, they would collaborate. They would ensure that constant hospitality, regular cooperation in a wide scientific circle, and planned travel to break the routine would answer their need for stimulation. But Fran, always independent, always driven to prove herself, and always supported by Hammy, would soon begin staking out her private territory.

Meanwhile their own wild place, isolated and full of things to learn, was waiting. They set out. First, they drove over much of the central and northern part of the state to collect information from the fall chicken kill. "Oaks and maples scarlet to maroon, aspen and white birch clear gold, some of the marshes already browning . . . blue lakes everywhere . . . a very fine state," Hammy wrote to his parents. But the results! "In spite of radio and newspaper publicity and personal letters to fifty hunters it was a dismal flop. Too much trouble for them to spend ten minutes for conservation's sake, once they had had their fun. We chased them up in the brush while the hunting was on and got some data that way, and the Forest Service had a few men out doing the same thing for us, so it wasn't as complete a fizzle as it would otherwise have been. Still, far from satisfactory."[28]

Leopold had found their base. His friend, Clyde Terrell, owner of Terrell's Aquatic Nurseries of Oshkosh, had a hunting camp on Sand Lake, a shallow pond north of Hancock, Wisconsin. Terrell warned that it wasn't much of a house, but the key was under a big rock at the left of the door, and they should use supplies that they found. They packed their car on Christmas Eve and went to a festive party in Madison. Then they drove the ninety slow miles on highways 51 and winding 22 to Hancock (population 235). They found the ragged cabin in a clearing coated with ice. Hammy shone the headlights on the door as Fran, in heels and silk stockings, felt around several stones, all firmly glued to the ground by ice. "We'll have to break in," he decided.

Inside, their hastily built fire poured acrid smoke through the frigid room. "Chimney's clogged," said Hammy laconically. The sagging iron bed, piled high with quilts, looked inviting, but mice had found the quilts before they did. Fran's subsequent letter to the senior Hamerstroms reported only that Christmas Day was "bright and beautiful." Hammy was more factual. He had just had a molar ("open to the core") pulled, and it was sore. He was "Sick as a dog," that Christmas Eve, and in the morning, "the damn car won't start." The house remained cold: Fran used an ax to get enough of the hamburger from the frozen lump they brought with them to cook for supper. Hammy would have to walk three miles to get a mechanic the next day. He reread his letter and added a reassuring sentence, "It's a strenuous life but a fascinating one."[29]

"Fascinating" was Hammy's word for problematic. Actually, visiting Wallace Byron Grange had left them with many questions.[30] Grange had studied the chicken population cycle, and building on Schorger's gleanings from old newspapers that showed well-defined lows in chicken numbers in 1857, 1867, 1878, 1887, 1897, 1907, 1918, and 1927; he added the years of 1937 and 1957. He saw "remarkable . . . regularity" in the cyclical pattern.[31] He estimated that Schmidt, working mostly alone, had banded 650 sharp-tails and 221 prairie chickens between 1931 and 1935, but he had been able to trace only 119 of those sharp-tails (and a mere 17 chickens) on booming grounds.[32] Fran and Hammy hoped to do better in the two winter trapping seasons and two spring booming seasons of their fieldwork time. (They aimed to finish their degrees by 1940.)

Challenges awaited them. For one thing, they had no appropriate traps. With Schmidt's plans gone, they had to reinvent the wheel. Their first effort, based on the design used for phlegmatic pheasants, injured these wildest of wild birds as they tried to escape by hurling themselves against the wire mesh. Courses in veterinary science at Ames hadn't prepared them to sew those seriously sliced scalps, so they killed the worst-looking birds, banded the others, and decided to try netting instead of wire. They had none, so in an endeavor that filled many evenings they made the netting. Fran reported, "I found instructions on page 397 of my grandmother's *Encyclopedia of Embroidery*. All those skills—lace making, embroidery, tatting—that I had labored painfully over with Fräulein ended up being very useful." Chickens might shun their nets, made from odds and ends of varied hue, so they dyed them, fixing the olive drab dye with salt. The netting on two such painstakingly furnished traps became food for salt-hungry deer, rabbits, and squirrels. They resorted to tip-top traps that sometimes injured the bird as it was tilted into a compartment.

They stayed at Terrell's for only one winter of banding chickens. It was too far from the marsh. It was simply too demanding to manage a consistent, efficient routine of locating traps where flocks congregated, baiting and checking them twice daily, and above all, unsetting them when the frequent storms came. No biologist wants to find frozen birds after a storm blows out, and it was after one freezing ten-hour day of unsetting traps after a unexpected blizzard that Hammy decided: "We can't do this another winter. It's just too chancy. We'll have to find a closer place."

That spring, while exploring for a place to fly his birds with falconer Bill Feeney, they drove west on County Trunk O toward Leola marsh. They passed a faint driveway leading through large trees to a ghostly gray house, stopped, turned, and drove in. Shingles were missing; a back room seemed barely attached to the house; plaster lath showed through holes in the walls. It was almost a ruin. "It must have been beautiful once," said Feeney. But, the Ware place, as it was called, was four miles closer to Leola marsh than Terrell's camp.

It took Feeney hours to find them the next fall. No one, he thought, would live in that wreck—which was exactly where he found them. They had fallen in love with the walnut grove and huge white pines around it. The owner was willing to let them live there; Leopold joined the shingling crew of Madison graduate students who came up one weekend to prepare them for the fall rains.

Leopold looked the Ware place over with an experienced eye. "Fran," he inquired, "is this a house or a camp?" He helped her dig the outhouse pit. Some layers of sand were dark ("Must have been a fire about then," said Aldo), some golden, water washed. "What do you suppose, Fran, grew in this area after glacial Lake Wisconsin retreated? It was cold in that lake bed." They leaned on their shovels for a minute before he went on. "I'm sure of one thing," he said. "People make outhouses much bigger than they need to be." She nodded, laughed, and drove her shovel again into the sand.[33]

Aunt Ruby, now old and alone, with plenty of time to write letters, recorded her admiring view of their life:

The young pair got a farmer to let them take possession of his abandoned house, in exchange for their putting it into livable condition, one year rent-free. They got odds and ends of lumber and material from a construction company, with their own hands, at no outgo, put in missing windows, frames and all, doors and frames, even steps to the door—and one missing wall—etc. They heat the place with an old-fashioned base-burner, bring water from a well, cook on a gasoline stove, use an old-fashioned galvanized bathtub brought in when needed—etc. and keep expenses down to about $30 a month,—living on savings from Ames and present stipend (twice a bonus of $600.00 has been a surprise-gift from the university to F—Jr.).[34]

They called their house "The Pines," appropriately enough since it stood in the very first pine plantation in Wisconsin. The original deep forests between the meadows of marsh grass and areas of high water had been a rich resource for early settlers. In the 1860s, a foresighted man would buy twenty acres of such land along with his eighty acres of sand to provide yearly hay crops and timber as needed.

Walter Ware was such a man. An energetic twenty-one-year-old New Englander, he moved to Hancock in 1856, married, and stayed for the next thirty-two years, raising a family of nine children.[35] He liked pine trees, patriotically planting 1,876 of them on his place and finishing on the centennial of the American Revolution. Neighbors scratched their heads. "What's going on out there? You're barely done grubbing that land, and now you're planting more trees."

"There ought to be some pines in this country," he said. "I'd rather look ahead than back on the 100th anniversary of this nation."[36]

In 1962 an interview with Ida Barnes, widow of the owner after Ware, told how her mother had helped plant those trees, in what was termed an "old field" only ten years after its clearing. Ware found a site rich with seedlings two miles east of Leola marsh.

[Here were] . . . open-grown, limby white and Norway pines. On slight rises seed from these trees had germinated and taken root. Because this soil was constantly moist and loose it was easy to pull the tree seedlings with little injury to the roots. . . . A yoke of oxen, a large "stone boat" . . . and a half-dozen youngsters barely in their teens . . . [transported] and planted the trees . . . Over . . . [which] the rear-guard flights of passenger pigeons . . . took place as adult birds built their nests on the broad, horizontal limbs of great pine trees in the marsh border, and took flight to the oak-studded terminal moraine 4 miles to the east. . . .

Planted trees had "closed in" and produced a soft, clean bed of needles when the steam dredges gashed long trenches in Leola marsh. . . . Drained of its waters . . . the marsh duff turned to tinder. Fires burned the very roots from the trees, some younger, some older, than the seedlings which had been rescued for the Ware plantation.

[The pigeons] witnessed the influx of prairie chicken as grass carpeted the often-burned marsh. They saw [their] . . . decline . . . to the point where Leola is one of the few last marshes in Wisconsin upon which the booming of the prairie chicken can be heard.[37]

Tree-loving Ware eventually moved south to Arkansas so that he could count on a dependable harvest from his apple trees. New owner

Shirley Barnes told Hammy that when he finally sold the place in the 1950s, the buyer had promised him that he would not cut the pines as long as he, Shirley, was alive. "It wasn't more than a few days after the deed was registered that the saws were at work in the plantation," reported Hammy. "It broke Shirley's heart." (Some maintain that a district forester advised harvesting the older trees for timber, "to improve the remaining stand.") There were still tall pines on the Ware place in 1990, but the house had fallen in to a cellar hole. Area residents could still, in 1975, see decaying shingles—the remains of the roof that Aldo Leopold helped put on.

They lived the history of the place. Across and down their road lived old-timers Mr. and Mrs. George Schofield, who remembered the passenger pigeons. Hammy and Fran visited the Schofields, then returned home to record the conversation in their notebooks. In 1975, they wrote up those notes for the summer issue of the *Passenger Pigeon*.

Mr. Schofield says some pigeons came through each year—enough for good shooting, but nothing like the quantities seen in the . . . the year of the big nesting. [That year] George Schofield had to sit in [the] grain field with dog and gun every morning till 10 o'clock . . . to protect seeding wheat (wheat sown on top of ground, dragged in) . . . [In spite of everything he could do, pigeons] spoiled crops that year.

. . . His father once fattened up something like 1,700 [squabs] in their barn. Mr. Schofield's mother salted down pigeon breasts and dried them like venison. The breasts were strung on a string to dry. These were daily taken to school for lunch with bread. . . .

I can almost hear old George Schofield talking—never saved the crop; we lost it ALL . . . the meat was so DRY, and it's just about all we ever had in our school lunch. . . . "Course fresh breasts tasted fine and the squabs was nice. . . . there was an awful lot of work for a boy them days—we lost the whole crop—the last year of the pigeons was TERRIBLE."[38]

They relished such accounts, but the marsh was even more rewarding. That expanse, no longer the original wilderness of tamarack, pine, and grass, was remote and relatively unspoiled. They used the hay camp shanties scattered over it as blinds. A local writer fondly remembered his farm boyhood on the Buena Vista marsh a few miles north and described what had been lost.

When the Ice Age receded twenty thousand years ago, the flood of all floods spilled beneath the fog-shrouded ice crags. The water ran faster than the Wisconsin River could haul it away, and . . . formed a five-million-acre lake, great and shallow and cold [stretching] . . . from south Plover to Bancroft, from the Buena Vista Hills to Wisconsin Rapids. It . . . served as the discharge area for the hill lands and all the surrounding area. The waters moving into it were nutrient rich, causing vegetation and animal life to flourish.

Being shallow, that water basin began to fill with used bits of the lives it fostered: tamarack, aspen, ferns, sphagnum moss and cattails. . . . The soil was a rich black muck with a peculiar sponginess to it. . . . The place was enchanted. Farm boys sent there to mow hay could watch hawks, families of hawks, circle to drop like bombs and rise with the wiggling rope of a snake locked in . . . their talons. There were countless nests of meadowlark, bobolinks, red wings, killdeer, finches, bluebirds, white throats, wrens, crows and hawks. It was a place of badger holes, skunks, coyotes, muskrat, deer, ground hogs, weasels and bear tracks.[39]

Now a flat expanse, bordered by the Little Pinery that reached north from the Wisconsin River almost to Black River Falls, appeared. Eastward lay a broad area of Plainfield silt loam. Islands in the former lake survived as local outcroppings, called mounds: Owens Rock in Leola, Mosquito Bluff to the east in Buena Vista, Rabbit Rock, and majestic Roche-a-Cri, site of a state park, near Adams.

The Leola and Buena Vista marshes covered much of the remaining ground. Native American hunters left arrowheads there, later found by settlers eager to exploit the rich resources available to them— bark for the tanneries; shingles for homes; wild hay for their cattle. Regular spring burning, controlled by marsh and wet woodland, kept hay meadows clean. Gradually what had been public land was sold to private owners, especially after dry years in the 1880s made plowing for row crops feasible.

No one then thought of weather cycles or habitat. To land-hungry settlers drainage seemed logical: districts could be formed to dig ditches, build small dams to control water levels, and level maintenance assessments. Trustees of Bradley Polytechnic Institute in Illinois saw a chance to create an endowment by draining and selling the marsh. Farmers, seeing the lush growth of wiregrass and marsh hay, figured that valuable crops would grow as well. Ditching went on from

1904 to 1911.[40] A Plainfield resident remembered his grandfather hauling coal with three sleighs and six horses. It fueled a floating dredge that weighed nearly thirty-five tons and ditched over the ice all winter. It cost $16 to ditch a mile.

It was the Necedah story all over again. A Wisconsin regional planning committee of the early 1930s diagnosed drainage as "a disaster that ultimately turned a thriving ecosystem into a dried-out, burned-over wasteland."

Fire could race across a marsh and burn one inch off the top of a thousand acres. It might linger long enough to incinerate two inches, or six. It could burn downwards in one spot or skip around and carve out yard-deep pond-sized potholes . . . . Already pocked and gouged by the wind, the sand country was now also scarred by craters where the peat had burned. Ditchbanks, where long rows of peat lay piled high to dry in the sun, were notorious for fires that smoldered and smoked for months at a time. Long afterward, Floyd Reid remembered a smoky summer's evening on the Leola marsh when he watched an orange sun set over pale green crops in a black marshland field ringed by ditchbank peat glowing red . . . by the 1920's, peat fires were common in spring, summer, and fall. Only a heavy covering of snow smothered the flames and made winter the single season of the year when the air . . . did not smell of burning peat.[41]

Wheat grew in the ashes—at least for a couple of years until the crop's heavy potash requirements depleted the natural supply. Then came corn. But, "Every five years we get a frost in July; every five years we get a frost in August; every ten years we get frosts in July *and* August" farmers reported.[42] Some went broke; others toughed it out.

They consulted professors of agriculture who recommended potatoes. The peat was wet; so growers had to put clogs on the workhorses to keep them from sinking too deeply into the sticky stuff.[43] In the nineteenth century, "taters" had made good money on newly cleared sandy acres for farmer and middleman alike, as imposing houses in Plainfield still testify. Now, late blight took the marsh fields; storage shelters full of potatoes dissolved into evil smelling liquids. "I watched a year's work just trickle out that shed door," said one resigned farmer.

Some Buena Vista landowners put the land back into pasture. Others contracted with seed companies for bluegrass seed. Local farmers grew the crop and stripped ripe seed heads from the standing grass on thousands of acres each July, bringing summer work for two hundred men and welcome payment for the crop. After the morning chores, men rode the tractor-drawn stripper and loaded sacks from the hoppers into trucks. Evenings, they went home to milking or land preparation after supper. The modest wages "sure come in handy," they said. Thus, by chance and circumstance, a substitute prairie, an almost ideal habitat for prairie chickens came to be, just when the species was fast disappearing elsewhere.

The whole of Plainfield Township was poor;[44] and politically the county was conservative. Waushara was the only Wisconsin county that went for Alf Landon in the 1936 New Deal landslide. Federal money had helped many farmers survive. Some of these now preferred to rail at "gov'mint." Federal funds paid half of the County Agent's salary, but one such man regularly and forcefully proclaimed the iniquity of taking federal grants and aid.

But what a place for chickens! In 1939, hunting prairie chickens was still a local sport, one the Hamerstroms enjoyed. Aldo often joined them at the Ware place.[45] Out early, they would flush a flock and then follow it, reading the subtle signs they knew. Fran and Hammy remembered coming home from a hunt with a meager result of a single chicken. They found Leopold, his gun leaning in a corner and two ruffed grouse, which he had bagged in a brief swing through the surrounding woods, "hanging by a thong on a nail nearby." Sometimes they would share a simple lunch and then hunt all afternoon. Over a glass of whiskey that evening they would compare observations. Aldo asked questions. "Why not count some of these winter flocks?"

Hammy, made confident by experience, responded firmly, "Aldo, it appears to me that winter is the worst possible time to count chickens. If we want to learn what is happening here, we must count during the spring booming."

Leopold invited Hammy and Fran to stay with his own family on Monday nights after the seminar that all graduate students attended. The three of them would then continue their intense shop talk—until Estella Leopold, the mother of five, grew weary of cooking the meals, making up their beds, and cleaning up for everyone. Childless Fran, brimming with enthusiasm for conservation and uninterested in homemaking, didn't think to share these mundane chores. (If there was

Aldo Leopold with two of his prize graduate students, probably in 1939. Hammy assumes an ironic professorial stance; Fran is her photogenic self.

coolness between Estella Leopold and Fran, it was aggravated by these practical concerns.)[46]

Back in Plainfield, Hammy found his predecessor's reputation to be a burden.

Franklin Schmidt was odd. He could not sleep indoors; when he came to see us, he slept outside. He lived in a farmhouse with his mother, and when it caught on fire, he went back in to save the papers and died. We survived eight years of the Franklin myth, the work he was doing, his seventeen papers in progress. Once Hammy said gloomily, after hearing for the dozenth time from a cooperative warden about the way Schmidt would have done it, "Did I ever have an idea that Franklin Schmidt hadn't already had?"

I lost my temper. "So he had seventeen papers, unpublished. If they were so good, why didn't he publish them? You've accomplished things, published things. I'm tired of hearing about Franklin Schmidt. That was that."[47]

With refreshing candor Fran confessed that had they read the existing literature more carefully, they would have learned that cocks go deliberately to a selected booming ground. Instead, they tried to scare the birds into view. Hammy, the observer, crouched in one of their rudimentary blinds while she flushed the chickens toward him. Hours of wading through popple woods to reach his observation point in a grassy opening might result in a few minutes of observation.

To trap, they tried bow nets, triggered by a string. Soon, in two locations, each was successful, each stowed the netted bird in a sock. Fran was jubilant as she went to meet a subdued Hammy. "My cock is dead, Fran," he said wearily. "How's yours?" Her sock, too, contained a dead prairie chicken. Was it shock? They never knew, but they handled birds very carefully from then on. Even more, they learned never to load more than five birds into one burlap bag because the bottom layer would smother.

To distinguish and remember individual birds Hammy suggested sketching distinctive marks. Not content to simply sketch the scars of puncture wounds on a bird being observed, he repeated the exercise after twenty minutes. He examined his sketches, then looked long and hard through the binoculars. He sighed. Then he grimly announced, "Those puncture wounds were ticks—just ticks—walking across the air sack."[48]

There was no research allowance; they cut popple poles to make the frames of traps and scavenged old nails to put them together. Their stipend ($50 a month) covered little more than their own modest expenses. They needed to get to the regular seminars in Madison to report on their progress and exchange ideas, yet sometimes they could hardly scrape up the $2 needed for gas for the 180-mile round-trip. Fran remembers these years as a hungry time. "We never drained the sardine oil. We drank it."

Their improvised blind for spring viewing was a card table covered, like a child's playhouse. They tried blankets, which flapped and spooked the birds; then burlap, which hid too little. Even the slightest movement caused the birds to flee. Finally, a scrounged refrigerator box gave them their first useful blind.

Presently Fran found her own thesis subject. Was it by chance that she picked chickadees? Perhaps she knew of the Leopold family's interest in chickadee 65290, a banded bird that appeared at the shack in 1937 and for five winters thereafter, and of Leopold's musings on that bird's "extraordinary capacity for living" that later appeared in the "December" chapter of *A Sand County Almanac*. Whatever the case, he greeted her thesis proposal with pleasure.

Fran chuckled about the ease with which she gathered material: "Hammy watched a resident flock out the window while he shaved in the morning. He called out observations and identifications to me, and I recorded them." Nonetheless, her curiosity and ingenuity led to the study. She asked the seminal question: was one bully responsible for the chickadee battles she saw outside the kitchen window? She caught that bird much as children catch chipmunks, with a baited box propped up with a string. She banded it with a gold link from a necklace; then devised colored bands from celluloid toys. These proved hard to see. Hammy thought of feather-marking the tails. Eventually she discovered that a once-defeated chickadee would be defeated again if he fought with the same bird.[49] Her master's thesis was "Dominance in Winter Flocks of Chickadees."[50] She was the only woman to earn a graduate degree with Leopold.

Alan, their first child, was born on 9 November 1940. His birth changed little for them. Accustomed to frugality, they simply rigged up a pulley and slung his basket from the ceiling of the drafty old house. Alan slept cozily through the night.[51] Their old friend Professor Joseph Hickey memorialized their child-rearing approach: "The week before Pearl Harbor we went up to Plainfield and survived the Hamerstrom treatment. There was a baby. The first one home from the field would change the baby. We all tried not to be the first one home."[52]

If motherhood changed Fran, it hardly changed her direction. Little talk of pregnancy, deliveries, or nursing, occupied her conversations. She, an over-directed child, made clear her intention to treat her children differently. She did speak of licking Alan's head shortly after he was born. Did she derive this uncommon "joy of motherhood" from animal behavior? It seems possible, though she might have said that it was pure instinct.

Well-meaning women friends in Madison (members, she says, of the "Life-will-never-be-the-same school of thought") declared that she couldn't stay "up there in the cold" with a baby on the way. She reported this opinion to Hammy. He paused and then queried, "What about the Eskimos?" Fran credited this matter-of-fact acceptance of pregnancy as a normal event with being, "one of the best things that ever happened to me."[53] The couple raised eyebrows by treating their two healthy children as tough little mammals who would thrive with minimal care. Richard Hunt, their first helper, testifies. "Alan hardly ever cried." But, "They were mighty offhand about the children," said Carolyn Errington.

The work went on, uninterrupted, and two theses came from it. Fran's thesis won a prestigious publication; Hammy's "A Study of Wisconsin Prairie Grouse: Breeding Habits, Winter Foods, Endoparasites, and Movements"[54] exemplifies graceful and attractive scientific writing. He arranged for copies to be sent to several people that he respected—Leopold, Grange, and Cox among them. Leopold's note of appreciation on 21 March added "miss you here;" Cox sent him an admiring letter on 6 May 1941.

He kept his eye out for employment, and hoping to direct fieldwork—preferably on chickens—he applied to be director of a Pittman-Robertson grouse study. Grange, more experienced and senior, was chosen.[55]

Then the position of curator opened at the Edwin S. George Reserve (part of the University of Michigan's Museum of Zoology). Hammy found, in the interview, that the position included research support for whatever expedition he chose—which would let him carry on with the spring prairie chicken count. So, with Fran's 1940 master's degree in hand, and Hammy's thesis well on its way, they packed up again and drove to Pickney, Michigan, a few miles north of Ann Arbor.

They liked it. They combined hunting in the Lower Peninsula with scouting for prairie chickens. They found smaller bunches of chickens in Michigan than in Wisconsin; sharp-tails were taking over in

Fran and five-month-old Alan in 1941 at "The Pines." Together, the Hamerstroms
planted one thousand pine seedlings to celebrate his birth, as they did later on
spring expeditions to observe the booming after Elva's birth in March 1943.

planted areas. Hammy's detailed report to the Michigan Conservation Department went to Leopold, too, with a scrawled note: "Southern Mich does'nt even have an open season . . . remnant flocks . . .
[are] comparable to Faville Grove. . . . This is Central Wis without
marsh—popple is on sand—plus Cretargus [Crataegus, or hawthorn]
in large stands plus large areas of orchard-like black and choke-
cherry."[56]

When Hammy announced his intent to go to Wisconsin for the expedition time allowed in his contract, his employer sputtered: "I
expected that you'd suggest South America, at the very least!" The
couple lived and worked in Michigan, but their hearts were still in the
sand counties. They looked forward to packing their gear and heading
back to Wisconsin or the prairie chicken count. The reunion with
friends and fellow workers, catching up with Aldo, working on the
count with his graduate students, and the return to the familiar wild
brought deep satisfaction. This was the life they wanted. In 1942

Hammy looked ahead: "Next year," he said, "with this pool of expe-rienced observers, we can staff a few more blinds."

It was not to be. Soon they, and most of the observers, would be far from the booming grounds. In 1943 Hammy enlisted. Few foresaw how long the interruption would last.

# 7

# An Interruption

## World War II

*Theirs is the hollow victory. They are deceived.*
*But you, my brother and my ghost, if you can go*
*Knowing that there is no reward, no certain use*
*In all your sacrifice, then honor is reprieved.*
—Herbert Reed, "To a Conscript of 1940"

Hammy, with one small child and one on the way, had watched the coming strife in Europe in deep concern. He had always been against involvement in war, as his letter from Iowa reporting a California Supreme Court judgment indicates. Two clergymen's sons were expelled from college for refusing to register for the draft; the court decided that even in land grant colleges, military training could not be forced upon conscientious objectors."[1] He had declared himself to be such an objector but after Pearl Harbor, when he discovered that that status required religious beliefs that he could not claim, he abruptly enlisted.[2] A recruiting officer had told him of the need for more men with degrees and implied that a doctorate would ensure a research assignment. That was not to be the case. Their son Alan, who was of draft age during the Vietnam War and may have talked with his father about duty and principle, believes that his enlistment caused a temporary rift in their happiness: "It was the low point in their marriage. She was against war, and he hadn't worked it out with her ahead of time."

Fran sniffed at this. "Alan was three—hardly old enough to remember those years." Then she wrote out her reaction as she remembered it: "I said, 'The first whiff of war you sign up to enlist. Can't you stand by your principles? You are being taken in.'" After a moment's reflection, she added: "Of course I followed him from base to base with our young children."

Putnam Flint, himself a decorated survivor of the European theater in World War II, summarized the situation. "It was a difficult time for a lot of people. I only saw my brother-in-law once during the war, but it is clear he was no soldier. He had his lieutenant's bar on the wrong side of his collar!"

Throughout their early years, Fran had accepted Hammy's decisions. When he had made up his mind, it did little good to argue, as she made plain in her autobiography.[3] Whatever upset the enlistment caused, her code, "Follow your man!" remained intact. When he was in basic training in 1943, Fran, in Michigan, cared for their new daughter, Elva, born in March, as she polished up the *Prairie Chicken Bulletin*, a new venture of Leopold's.[4] She joined Hammy briefly at Randolph Field, Texas, for part of that time and then returned to Michigan; he came home in July, as a second lieutenant. Her feelings showed: "Hammy was of those 'one-day wonders,' as the career sergeants called the officers who got their bars in a single day. The only salute he really knew was the Boy Scout salute. Ann Arbor was then a huge training center for the Army, Navy, and Air Corps. On a walk downtown, he didn't know whom to salute and who would salute him first. Peering intensely into a shop window when a uniform approached saved him."

Yet a summer letter from Randolph Field, the "West Point of the Air" according to the letterhead, displays a warm, natural interaction. He addressed her as "Lovely," gave her several errands, reported a decision to buy tennis racquets in Texas, cheaper there than restringing in Ann Arbor. He spoke of the likelihood of transfer, and the heat. He wrote an illustrated note for Alan. And he said, "I've already told you . . . how delicious iced tea is here, and how often it is necessary to drink. Whenever I do, because it is so pleasant I think particularly of you. And whenever the breeze comes in my window at night—a blessed relief— I think of you coming in and running your fingers through my hair. Soon, I think, we shall be together and meanwhile, these poor substitutes help to make life more bearable without you. I love you."[5]

Wartime: a tense Hammy in uniform in 1945, on a night out with his wife.

So, after a temporary posting in Nebraska, he went to Idaho. Fran loaded the essentials into the trailer. With Alan beside her and Elva in a bassinet in the rear seat, she pulled the trailer to Idaho. They slept in the car. On arrival at Mountain Home Air Force Base, where wives were not allowed to stay on base with their husbands, they proceeded to find a two-room shanty on the Snake River some miles away. Hammy sent her to Boise to buy a refrigerator; she came home with a duck boat instead. "We hung meat outside on cold nights, and wrapped it in a sleeping bag in the morning to keep it cold."

His contribution to the round-robin letter that Leopold's students maintained during the war was self contained. "Howdy!" he began:

Very busy here lately—Today is my first day off in 3+ weeks. Among all the assorted land-and-sea-going brass in this outfit, Art and I will have to keep our right elbows oiled up and pronounce our "sir's" trippingly on the tongue. On this point, one bright saying which I can't resist (after which—word of honor, no more): Fran to Alan "That's Lieutenant Hammy." Alan, after howls of laughter, "That's just Hammy in lieutenant's clothes."

For the present I am a professor of oxygen—strictly a teaching job, no research, and no more than a nodding acquaintance with airplanes. Not what I had expected . . . by a long shot. But there it is. One other officer is in our outfit—1st lieut. very good egg—and twenty-five men, a small building, a small tractor-trailer pressure chamber which sits outdoors and freezes up in the weather. . . . I learned a new word on joining the Army—snaphoo—and I'm learning the finer points of its definition daily.[6]

"There it is." His mocking phrase, "professor of oxygen," and the mild "Not what I had expected . . . "were as much complaint as Hammy allowed himself. He preferred to tell his friends about, "the best duck hunting we have ever struck," their bird collecting, and their plans to go after mammals to take advantage of the country—"so different from any we've ever worked in before." He signed off with a sardonic "for the present, *a dios.*"

In duck season, Fran remembered rising early, leaving the sleeping children, and heading upriver to their mooring spot. As ducks rose, she shot; then moored the boat and walked home, returning to still sleeping children and pumping water or dipping it from a hot spring to wash diapers. Hammy floated his way home for splendid shooting during the evening flight.

When his unit was posted to Harding Field Air Force Base in Baton Rouge, Louisiana, Hammy—a popular officer—worried about, "the men having to move that heavy pressure chamber that had to be horsed onto a railroad car." He stayed with them, through a major sleet storm, until the task was done. "They loved him for that," Fran recalled.

Housing was almost impossible to find in Louisiana, but her natural optimism and high spirits stood them in good stead. They soon gave up on classified advertisements and realtors, and used Hammy's free time to drive around the countryside, searching. On one such excursion, she told of a sudden find. "Stop! Look, darling," she pointed, "look at that huge plantation house, up that long drive. It's just there, behind those wonderful big live oaks. Look at that Spanish moss!"

He slowed, peered. "Somebody's living there," he replied flatly.

"Maybe they're moving!"

Hammy put the car in gear and moved down the road.

"Darling, please turn around. Let's go back and ask! Please, Hammy."

"Fran," his tone was patient, "I saw clothes on a line, and people on the veranda."

"Please! Please, Hammy. Let me talk to them!" He turned around. They drove down the curving driveway. Fran ran up the steps. After a few moments she was back.

"They're not moving till tomorrow. Then we can have the house for $12 a week. And," she smiled triumphantly, "we can have servants for a dollar a week more. They belong to the house." However, it was too good to last. In January Hammy was promoted to First Aviation Physiologist and Staff Personal Equipment Officer, at Biggs Field in El Paso, Texas. They packed up and moved again. Fran became a phlebotomist at Beaumont General, then the largest army hospital in the world. They liked the Mexican food and their house near the Rio Grande, on which they could hunt ducks in season.

Hammy's work was frustrating. His deep feelings that his skills, training, and capabilities were wasted came through when he told Fran, "Any high school teacher could do what I am doing," and added, bitterly, "and probably do it better." As staff in the Altitude Teaching Unit of the Air Force School of Aviation Medicine, he managed the equipment of the bomber crews and trained fliers in safety and efficiency. He lectured on parachutes (two hours); oxygen, emergency equipment, ditching, and flying clothing (one hour each); on drills for various ditching options, including water (in the pool, two hours). He showed such training films as *Night Vision for Airmen* and *Land and Live in the Jungle*. His lectures, models of clarity and reality, were dotted with pithy statements of the things airmen needed to remember.

The harness must be fitted: snug in a sitting position, definitely too tight to stand in. A loose harness may dump you out when your chute opens, or spoil your post-war family plans.

In free fall, you are a remarkably difficult target; floating down with your chute open, a sitting duck. It takes about 20 minutes to get from 40,000 to 5,000 feet with your chute open, about 2½ minutes in free fall.

5,000 feet is the best place to open [parachutes] . . . it is where the horizon suddenly widens out and the ground seems to jump at you.[7]

There was no way to kid himself about his assignment. He knew what faced the men he was training. "Poor devils!" he thought, looking out over the young faces, headed for combat in B-29s. "Better trained than untrained, I suppose." Such mind-set, which he kept to himself, found voice in another military instructor's moving lines,

Of Van Wettering I speak, and Averill,
Names on a list, whose faces I do not recall
But they are gone to early death, who late in school
Distinguished the belt feed lever from the belt holding pawl.[8]

His responsibilities as equipment officer were tedious, but he was as effective as he could be. In August 1945, a time of demobilization and rapid personnel turnover, he was criticized for deficiencies. He responded stoutly: "The bins . . . have been repeatedly . . . approved by the Air Inspector, Technical of this station . . . . Napthalene flakes are being used, but because of the open construction of the lockers and bins, the fumes are quickly dissipated. . . . There have been four supply officers in this Department since 24 May, three of whom held the job for less than three weeks each . . . [this unit] has been consistently hamstrung by inadequate manning. It is the desire of this Headquarters that Commanding Officers use greater care in the selection of Personal Equipment Officers . . . assuring . . . [that they be] interested in, and well suited for, such a position."[9]

A mishap in May brought his sense of justice to the fore. During one hastily scheduled night exercise, the life-raft compartments on two aircraft opened in flight. First Lieutenant Pablo Seguro was charged with maintaining liaison with Maintenance and Supply to insure proper stowage of life rafts. Seguro was the one the Army chose to blame; he maintained that he had supervised the task correctly. Hammy agreed. He had watched him load the first plane very late one night. Fran reported:

They were going to court martial Seguro. Injustice made Hammy's blood boil. He went to the Colonel.
"This action is inappropriate without an investigation."
"No investigation will be made." Hammy began interviewing the men. The Colonel told him to leave the matter alone. "If you don't, I'll have you court-martialed too."
I drove through the gate to pick him up that day. The guard motioned me

over, and whispered, "The lieutenant is really mad today!" But Hammy persisted and saved Seguro.

One of the very few service-related things that Hammy kept was his meticulously researched, nine-page, single-spaced report on the incident. It detailed the hasty notice for the proposed exercise and its timing in the middle of the night. He precisely described the loading he had observed. He listed his inquiries: at nearby bases he found records of seven recent, similar accidents. Reports concluded that the plane's flexing had caused the latches to disengage. He suggested that material failure be stated as the problem. There was, in the end, no court martial.

Such small successes gave him satisfaction, as had "preventing the worst mistakes" back in Necedah. Finally, that November, his replacement arrived.[10] He signed his final property turn-in slip—for his government-issue bicycle—with a small flourish on 21 December 1945.

The return to Michigan could wait; rest and renewal were priorities. "Why not go see Paracutin?" suggested Fran. "It's not every day you can see a new volcano. Posey will keep the children." She knew that Posey loved "Treasure" and "Blessing," her pet names for the grandchildren, and would welcome the joy they brought to a Christmas season darkened by Mr. Flint's final illness.

Elva vaguely remembers a "Hispanic mother" (probably a childsitter) who fixed her hair in elaborate braids decked with ribbon. Elva also remembers a crowded railroad platform and her parents' sudden, startling words: "We'll see you soon, darlings." An affable sergeant ("They had never seen him before in their lives!" said Putnam indignantly) agreed to deliver the children to Posey, who would meet their train. That was the last the Boston family heard of Fran and Hammy for some time. Christmas went unmarked, as did Elva's March birthday. Putnam says that his father finally asked his FBI connections to trace the missing pair only to hear, the next day, that they were on their way home. Sometime in this period, Leopold wrote, offering Hammy a teaching position in a department at Madison, Wisconsin, but as Fran reported, "We had commitments elsewhere."

This, the first of many such winter seasons in Mexico later, is puzzling. They certainly were in Paracutin, Mexico: there is a February letter to Milton from Fran with her sketch of the volcano.[11] They often disappeared on trips with neither notice nor reports on their return. But the vivid memories of Dagmar Lorenz (Konrad Lorenz's daughter), who stayed with the family in 1958, add uncertainty to the matter. "Fran told us about Mexico," she reported, "and said that she feared for Hammy's life, but that she cured him." Now a practicing psychologist, she said explicitly, "You must find out about that trip. I've always wondered why Hammy was always so silent."[12] And Os Mattson, who worked again with the Hamerstroms on their eventual return to Wisconsin, reported that Fran had told him of her grave concern about Hammy in that exact phrase—she feared for his life.

Neither Fran nor any family member provided clarification. An old friend spoke vaguely of some "serious illness." Alan dismissed such accounts as exaggerations or an attempt to explain a puzzling period. Elva's comment was simple: "Hammy was her rock. I can't imagine any kind of breakdown. Their expedition in Mexico could have been a time of healing for both of them. Fran was fearful, all that time. She told me how she checked his arms for traces of immunization—the revealing sign of a coming posting overseas—each night." Sending children, the younger not yet three, across the country with only a hastily arranged kind of safeguard might indicate some urgent need on a parent's part, but Fran and Hammy were unusual parents in an unusual situation.

The matter remains unclear. Hammy hated war; he considered the Vietnam conflict to be wrong-headed, and the army's use of defoliants and napalm to be "shocking" and "disgraceful." He readily agreed to be a witness for our eighteen-year-old son's application for conscientious objector status. Though he knew the boy only slightly, he drove twenty icy, dark miles to the Wautoma Courthouse for the hearing— the first such appeal in the history of our county. It was granted. The best indicator of his beliefs, however, was his involvement in European relief shortly after his resumption of civilian life. Binding up the wounds of war is magnanimous, a trait characteristic of Frederick Hamerstrom.

Besides, Hammy had been purposive and serious for fifteen years. He would continue to be seriously purposive, though perhaps a bit more silent, for many more. His years in the service and his judgments and the actions he had undertaken there furthered his organizational and supervisory skills. His persistence and Fran's loyal support are the vital traits displayed and strengthened in these years of interruption and growing depth—depth that back in Michigan would be called into play in what Fran termed "the most demanding adventure of our lives."

They looked forward to a return to their normal life as they made their way back to Michigan via Milton. They wondered how the Reserve had fared during their absence and happily anticipated resuming observations on the prairie chickens in Wisconsin during their annual spring leave. Their lives had often brought the unexpected. The coming period was to be no exception.

# 8

# The Action

## Postwar Scientific Solidarity

*We cannot love issues, but we can love people, and the love of people reveals to us the way to deal with issues.*

—Henri Nouwen

After the war, thousands of Americans responded to the desperate circumstances of German and other European civilians. The Hamerstroms brought special skills and sensitivities to that task. In December 1946 a letter from Joe Hickey and naturalist Margaret Morse Nice passed on troubling news from Ernst Mayr, Harvard's famous evolutionary biologist. European ornithologists were starving. Fuel, homes, and jobs were gone; cold and malnutrition had replaced bombs and armies as the enemy. Could American colleagues help? The members of the American Ornithological Union (AOU) decided to focus on sharing scientific knowledge along with material aid. They agreed that no potential recipient's past politics were to be questioned, and put appeals in the *Auk* and the *Wilson Bulletin*.[1]

Fran's ties to Germany were deep; her participation came from her heart. In 1945 and 1946 she received touching letters from her former governess who had been bombed out from retirement quarters in Kaiserswerth; her savings, in England, America, and Dresden were out of her reach; she was living with, and dependent on, a nephew in Düsseldorf. "Well we just have to bear it as a punishment from God, the ones who hated the N[azis] just like the N[azis] all the same." Fräuta blamed her troubles, and those of Germany on "so crazy Nazis . . . but

I assure you there were really not as many N. as your Government thinks more than half just out of fright as they made short process with anyone who did not do as they wanted . . . a terrible time thank God it is over." In her tidy German script she recorded happy memories, "My dear Frances . . . I see you with the conserve glass in your arm and at the pond scooping up some mess a treasure for you it had all kinds of creatures swimming in it you looked gleeful and how you loved to sing all those german songs, you were a joy to me."[2]

Fran and Hammy, idealistic as ever, became central to the effort, putting their habitual energy, ingenuity, and parsimony to work. Hammy, the team's compass and anchor, provided practical and ethical guidelines. Fran did the daily arranging and executing. Hammy said, "Ten dollars per CARE package will certainly limit the number sent," so she fashioned parcels from what they could spare and matched donors to like families so that used clothing would be utilized. It was satisfying to find "a tall American gull expert with a slightly older daughter whose clothes could be passed down to the child of a tall European gull expert."

On 23 January 1947, Fran wrote the first of dozens of letters. To Gustav Kramer, a German biologist living in Heidelberg, she wrote in English: " . . . how old [are] the children . . . are they boys or girls . . . send tracings of their feet (not of their shoes), perhaps we will find that we have some shoes of the right size."[3] To Erwin Stresemann, then the world's best-known ornithologist, she described their work: "I am working on a paper on the food of the Great Horned owl and my husband is gathering material on food of hawks, so . . . [we] need Uttendorfer's [sic] book on food of European raptors. . . . tell me where I can send for a copy."[4] Those first letters went by sea but as continued frantic appeals continued, they turned to airmail and her second letter got to the Kramers before the first.

Kramer and his Italian wife, Neni, spent the war years in Naples. Drafted, he served as "a horrified private" in the German army. Now he and his family were in Heidelberg, living on the family vineyard with his parents. Thus they had food.[5]

Paper was scarce; Kramer used the back of her first letter to answer. "When your airmail letter arrived, my wife suggested it was written by the infant Jesus Christ (anyway, it dropped from the heavens!)" How

Gustav Kramer shortly after the war. He was one of the main German facilitators
of the Action. This effort dominated the Hamerstroms' life for close to three years
and brought American scientists into lifesaving and creative contacts with their
counterparts in war-torn Europe.

interesting was chicken courtship behavior! He knew of similar behavior in male stickleback fish. A paper by Konrad Lorenz "would be
the reading's worth for you." Strangers had sent CARE packages; old
clothes could be remade for trousers for their sons, Peter and Lorenz.
But shoes! stockings! Those problems were "completely unsolved."
The active six- and seven-year-olds couldn't go out when it rained. Professor Von Holst was desperate for food and clothing. "He must not
know this act of witness of mine, of course." To repay them, perhaps
he could do bibliographical research? Or contribute German stamps to
an American philatelist? His own work on lizards and inborn reactions
to predators was on hold and his reprints were in Naples.

The tone of these letters—personal, tactful, informative—characterizes what they all called the "Action." In return, the Hamerstroms
expressed appreciation of the modest requests. "We are deeply moved
by the people who write, 'send no more, help others.' You asked what
Europeans can do to pay back: they can write about their work and
themselves."[6] Hundreds of letters ensued. Kramer wrote fifty letters to
find needy German scientists. Mail was slow; three intended recipients
died before packages reached them. One was Fräuta, Fran's "beloved
governess."[7]

Fran, who wrote letters even during the demanding periods at Terrell's camp,[8] urged prompt replies. Professor Mayr characterizes their
efforts as "heroic."[9] Modest Mrs. Mayr credited Fran and Hammy
with dipping deep into savings to fund their part of it. In retrospect,
she wrote: "As to AOU [American Ornithologists Union ] relief effort
. . . we have no records but vivid memories of a very lively activity. The
driving force without doubt was Frances Hamerstrom. The motto 'We
ornithologists who feed the birds in winter will not let our colleagues
starve in their time of need.' . . . Letters poured in by the thousands.
My major contribution was to translate and interpret these . . . a special request would be fulfilled by whoever heard about it and could
spare the item . . . for instance . . . a blanket for a displaced person living in a shack. . . . Some requests (as for inst. for an umbrella or dress
gloves) were received with less sympathy."[10]

Gretel Mayr mailed more than a thousand packages herself, paying $10.00 each for the CARE packages, and sorting and packaging
clothing she had begged from neighbors and friends. It was a complex

task—the wife of an American entomologist, for example, wanted to help a Heinrich Prell. Mrs. Mayr wrote Fran:

Please send me Prell's letter back. Don't expect too much, they are just a young couple with a beginner's salary. As in so many cases the kindness of the heart is not with the people who have the dough. . . .

I was worried about Jacobi, Schoewetter and Uettendoerfer. What shall these aged people do in all the time that elapses till word has come back that they could not travel to Berlin? . . . . Stresemann [may] . . . find some ways and means to get the package to them . . . . By the way, if Prell should have gotten a coat already, it will still be time if you let me know by airmail. The coat has to go to the cleaner first. I could add a suitjacket and a vest some non-ornithologist gave me. But the pants are missing . . . . I expect to hear from you about Goethe's family and Kattinger. The packages stand there, ready to be finished and mailed. I hope the shipping strike is over soon. [11]

Then Dr. Mayr got a request for a dark suit that a German could wear to give lectures, and thus make a little money. She packed up and mailed her husband's one dark suit. Months later, one Friday afternoon, he was asked to be an honorary pallbearer. That evening he asked her, "Where's my dark suit? I'll have to wear it."

"But I sent it to Professor Drost!"

"No! How can I get another?"

She hadn't enough cash. Banks did not open on Friday night or Saturday in the 1940s, and professors' salaries were small. They scrambled to get money for what was, for them, a serious investment. [12] Other cooperators like Ellen Ammann (who with her husband, Andy, had known the Hamerstroms in Iowa), canned and sold beans for postage money; others sent their children's shoes. "They can go barefoot this summer. They'll need new shoes this fall anyway."

Hammy and Fran discussed the mail each morning over coffee. Requests for CARE packages went to John and Jinny Emlen, Madison friends and dancing partners, who managed the Midwestern CARE donations. Colonel Irwin, Chief of Branch Higher Institutions, U.S. Army, Heidelberg, rushed Kramer's packet of requests through the army mail system. [13] Names came so fast that Fran wrote hastily telling Kramer to cancel a planned announcement in the *Ornotholgische Berichte*.

Some opposed the policy to question no one's politics, believing, apparently, that one should not send help to former enemies. Hammy and his committee stood firm. "Hammy risked his job to uphold the principles of the Action," Fran reported, still incensed at 80.[14] "We couldn't ignore some of the letters." She recited examples: "My daughter is five years old, with blond hair like floating flax. She is pale and no longer plays. She is ill. The cause is malnutrition." One couple begged for boots. "We request a pair of stout boots—size 43. We have one pair of boots so we must take turns: I work at night and he works in the day. We must spend most of the rest of our time in bed for warmth."

Day after day came such appeals. Fifty years and many degrees removed from those days, it is hard to select among those unexpectedly found after Fran's death. (She had reported that, exhausted, she had simply thrown everything away.) A widow wrote directly, begging for help for herself—underwear, warm stockings and a warm dress; shoes and clothing for her crippled son, and shoes, underwear, and outdoor clothing for a three-year-old daughter. They all needed food.

Responses to such pleas came from the Midwest and East, from Canada, the West Coast, even the Hawaiian Islands. By August, Fran's file contained a thousand cards. Hammy, occupied with what was a not entirely satisfactory situation at the Reserve, wrote the Kramers: "Although the exchange of letters has been between you and Frances (we'll have to explain that arrangement, some day), I have read your letters and feel that I know you, too. Thank you so very much for the help you have given us—we would have had a most difficult time without it. I hope you will never feel that you can't write us directly—a few days or a few weeks saved may make a great difference in how quickly we can find someone to take care of a new family."[15]

Kramer felt beholden; he wrote of possessions, "which may be called sterile in the present days." People would be happy to send those things to be "liquefied" in the United States as barter for American aid. He had some ancient engravings; Von Holst had an old Chinese shawl, some valuable violins.[16] The Hamerstroms were firm in their response:

Frederick and I agree that it is very generous and touching for anyone to want to send valuables, but that we hope you will not let anyone do it. The original concept of the Action might become utterly distorted. . . . We feel this very deeply. If you want to "ventilate" this matter farther, we will send your letter to the other committee members. Some of the donors are making sacrifices; I can think of none of these who could bear the idea of getting valuable presents. I know of no donor who is in real need. They have enough to eat and to wear. They will not be cold, and if they prefer to be shabby, it may be a good choice and they have the right to it.

Small presents . . . like Christmas and Easter cards, some little thing to amuse a child, might give great pleasure. Please note. The Hamerstrom family does not want presents. We get thanked in so many ways for our part in this. There are several who hold misconceptions: i.e., that we started the Action. We didn't, it was M. M. Nice and J. J. Hickey.[17]

The letter, written by Fran, almost perfectly illustrates the melding of their distinct approaches. The concepts in the first paragraph are almost certainly Hammy's. The language ("utterly distorted," "ventilate") is his, as is the care in crediting. The practical, earthy sentence about the right to be shabby is the kind of talk often heard from Fran. The combination is *echt* Hamerstrom.

A paper shortage in Germany kept Kramer from printing a journal he edited. Postal regulations in the U.S. defining length and girth required that large sheets of printing paper be cut, but printing cut sheets would increase Kramer's costs. Hammy decided that they must mail full sheets, in spite of worrying that rolling the packages would make the paper unfit to print. The rolling and tying of the resistant bundles "drew blood" from their hands, but those packages enabled Kramer to print his journal in Heidelberg.

Fran described the difficult situations and judgments that faced them. Few in this country realized that to hold any job in Germany— mailman, teacher, museum curator—one was automatically listed as a member of the Nazi party. The American anger toward the Nazis meant that, for one period, the Hamerstroms sent only to those identified as Nazis. Nobody else would help them.

Other complications were harder to understand. The articulate were easy to get help for, and one German who had written a good, heart-breaking letter, was promptly adopted, with food and clothing promised for the whole family. Yet, months later they learned that one of the children had died. Fran asked the donor how the adoption was coming along, but the reply came: "We don't send to Jews." Royalty also presented problems; they sent packages to a king because they, according to Fran, couldn't get anyone else to do so.

The Hamerstroms had not expected prejudice or problems of determining right action or blame. According to a child of the times, Gustav Kramer's daughter Elizabeth: "Fran and Hammy knew that everybody needed help, especially the children. They needed a definition of nazis and 'good people' because they knew that people prefer to send to decent people . . . respected by their European colleagues."[18]

With need outstripping donors, they sent out scores of mimeographed postcards; the Emlens arranged a radio program about Nikko Tinbergen and other well-known scholars now on the list. Hammy wondered, "Do you think it better to get a little help to many European ornithologists or to send more things to a smaller number? . . . How can we divide to best advantage the things that we can get?"[19] Said Kramer tersely: "a little to many, please, for the sake of morale." He went on to speak truths no less telling for their simplicity: "A lot of Germans, especially chased [displaced] ones, are vexed by idea that the whole world has no other aim than strangulating the German population. The . . . existence of helping souls abroad is a most efficient remedy; I had striking proves [*sic*] of that."[20]

Remembering their own student years, the Hamerstroms asked if students and amateurs were worthy.[21] Letters went back and forth, between Pickney and the Mayrs in New Jersey, Emlens at Madison, and the European helpers; they arrived at a working definition of appropriate recipients. Kramer, like the American donors, thought almost everyone was needy.

Just the moral effect of feeling the intention to help is invaluable. . . . Fritz Winziger, factotum of the *Vogelwarte Ressitten* for, say, 30 years, now chased from his place, an old man in a strange, overcrowded environment, was proposed by Schutz. I found it was justified. Stresemann hesitates, as he does in several cases of serious bird banders to whom the ornithological stations are enormously indebted. Had they not done their co-work, many questions of bird migrations would still rest at a more primordial level.

[Please pass] all requests through the hands of Stresemann. . . . the difference in our views will have disappeared . . . our cooperation will be perfect. He knows people much better than I do. As a matter of fact, he dropped out one man who had better stay out, as I agreed in this case. Do not fear your work will be late! Misery and starvation in Germany will not stop before 2 3 years.[22]

Hammy handled the case of Dr. K. in Schleswig-Holstein, who had "behaved badly." He phrased his letter carefully: "It is my impression that your name was deliberately withheld from those who were to receive help from the Relief Committee for Ornithologists. . . . [We] accept only those names which have been recommended to us by a group of our colleagues in Europe. If you will send me recommendations signed by the three following ornithologists (Stresemann, Kramer, and Ernst Schuz [sic]) asking that your name be included on our list, I will include it." Unsettled, he added a note on Kramer's copy of the letter. "Mrs. Hamerstrom and I don't quite know what to do in this case. We have been told that K. has behaved very badly, and, in fairness to others, should not be helped; for that reason his name has never 'officially' reached us. However, he has written us direct. Since we know nothing about him at first hand, it seems best to leave the decision to you. If he has your respect, we will be glad to help him." Still not satisfied, he added, "I do hope that Fran and I have done the right thing here."[23]

Fran kept track of requests and responses. She tried various systems; sometimes leaving letters in their envelopes and jotting notes on the back, sometimes typing strips of information to affix to each letter. Cartons of envelopes, some stored by date of receipt, stood on their floor.

The strips are eloquent. "Prof. Dr. A. Jacobi Hoe Str. 102, Dresden 27, Germany Russ. Zone lost everything, completely starved 77 years old, underwear, handkerchiefs, shoes size 9." (He got ten packages from six donors.) Kurt Walther, Rohrbacher Str. 135 Heidelberg, "is hungry and in rags."

The effort widened to scientists, such as mammologists, in other disciplines.[24] The process became more exacting. Some requests came without the necessary British, American, French, or Russian zone for the address. A few Europeans assumed that rich Americans could send luxury items. Fran told them of donors on modest salaries who sacrificed to pay the parcel postage. Some Americans, naive about malnutrition, were upset by such requests as the one for food from a short man with a forty-four-inch waist. Fran educated them: starvation creates pot bellies. She had to soothe a donor who resented a request for food after she had sent a powder-blue tweed suit, overcoat, and matching hat. "Conditions are very bad in Berlin. Frl. M. has a way with cats. She entices them to come to her; strangles them and slips them into her muff so she can later cook them."

Her Yankee resourcefulness came to the fore. She sorted unusable clothing—torn, dirty pants with the zippers ripped out—from the thoughtful, "pressed, carefully folded garments wrapped in tissue paper, with presents in the pockets for the children." She added things of her own: "a ski suit that had belonged to my father, thread, needles, candles, soap, a syrup can of home-rendered deer fat, and two Hershey bars." Sometimes she compensated for condition, as in Frederick's very bad boots. "We sent leather and thread for mending and coffee to drink (or if necessary to pay the bill)." She always shopped for bargains. Candles in the shape of black witches, pumpkins, and Christmas trees went to Europe after postholiday sales.

The Kramers kept Fran up-to-date. Was instant, ground, or bean coffee best? Who was expecting babies? Had some stockings been of very poor quality? Who had gotten them? She revised her request list as times changed. At first, bombed-out windows needed "that translucent material that farmers use on chicken houses." Later, "candy for the children, and slip a package of cigarettes into the toes of the shoes" seemed fitting. That sentence unfortunately followed a list of countries that forbade shipping cigarettes and brought a challenge at the annual meeting. "It appears, Mrs. Hamerstrom, that you have been asking people to smuggle cigarettes in shoes." She thought quickly, then spoke sincerely. "I knew they wouldn't get squashed there." Simple packing instructions could hardly be criticized.

On they went, pace unabated. On 20 February 1948 she wrote Kramer, "Personal packages to the following 17 people: 5 Kramers, 3 v Holsts, 4 in Lambert Schneider's family, Hilda Zimmer, Klaus Schmidt, E. Handke, Klaus Dylla, U. St. Paul."

Financing the expanding operation needed a dramatic effort. She conceived of an auction of paintings by European wildlife artists. The proceeds would buy food. Remarkable artists like Franz Murr, Rein Stuurman, and Otto Natorp responded. Fran's elated Christmas letters described a Wilson Club dinner meeting and auction at which they raised almost a thousand dollars. "Hammy bought three water colors!"[25] Eventually she managed sales of paintings for nine European wildlife artists, collecting payments (occasionally with some difficulty) and maintaining a bank account for each artist.[26]

The Kramers sent the Hamerstroms a painting at Christmas. Hammy wrote at once, though still a bit formally, to "Dear Dr. Kramer: This is plainly a picture that you are very fond of—for that reason we shall treasure it doubly. We framed it and hung it at once, in the children's room, as you directed."[27] Fran—well aware of German reserve—now felt free to write, "We would be very happy if you would call us by our first names. Hammy pronounced Hammi in German and Fran pronounced Frahn."[28] The idea of visiting each other surfaced and soon they were "Dear Neni! Dear Gustav!" She said, "I had a panic when you said you were 'resigning' and then to top it off added 'thank you'! No! no! no! you are *resigned*. Good. . . . Lorenz [the well-known scientist who had been in a Russian POW camp] is back—thank you for sending the news. We are packing up to go to the booming grounds . . . in Wisconsin. We . . . go back to our old house and must take a lot of equipment."[29]

Then she regaled them with her latest verse, inspired by her recent attempts to imp (particularize) male prairie chickens by gluing in colored tail feathers.

> The prairie hen will wonder soon
> But not because her love goes Boom
> Consider with what joy she'll hail
> The colored feathers in his tail.[30]

Shoes dominated months of the effort. Fran had one thousand foot tracings; she arranged a shoe offer—for a dollar she would send shoes to a European ornithologist in the donor's name. She acknowledged each dollar. A shoe store gave her boxes of rummage; repair shops donated unclaimed shoes and boots. She shined the old shoes and put in new laces. "Our children got a strange impression of how to treat shoes. New shoes were subject to duty; used ones were not. Alan and Elva spent many hours scuffing new shoes on the cement floor of our basement."

Konrad Lorenz, noted Austrian biologist in 1956. He became a Hamerstrom family friend and a model for Fran. She was significantly influenced by the simply and compellingly stated scientific understandings of animal behavior that he brought to a captivated public. Photograph by Elva H. Paulson.

By May 1948 Fran asked for urgent requests only. "We have to slow down a little. We were overrun for a while and quite frantic."[31] By August, they were able to concentrate on food packages to those without work and residents in Berlin and the Russian zone.

And when Hammy bought her a frivolous pair of dancing slippers, with high heels, built up soles, and slender straps for Christmas, Alan asked if he should scuff them. "No, dear," she said. It was then she realized—suddenly, overwhelmingly—it was over. It had been costly—in time, in energy, and in mental strain. In retrospect, Fran at age eighty-five described her own condition.

I simply threw away . . . all those shoe tracings, letters, and the card file. I tried to cook lunch—I just couldn't cook. Elva wanted a dress made for her doll. I got out some material and couldn't cut it out. I couldn't seem to shake that mood.

Hammy saw the situation. He went and got someone to take care of the children and said, "Come on, Fran. We're going to look for cranes."

Out there, in the marshes, I said, suddenly, "They're coming!" Hammy looked at me intently. "Where, Fran?" he asked.

"I can hear them!" I cried. I've always had worse ears than any of the rest of the family.

"I can't hear anything," he said.

"Listen! They're getting louder and louder!"

After two or three minutes, he put up his binoculars. "Good God! they are coming!" I never had that kind of hearing again in my life. But I began to recover my balance. Hammy knew being outside would help.

By the AOU's formal count, more than three thousand parcels (849 of them CARE packages) had gone to scientists in Germany, Austria, Hungary, Poland, Finland, England, France, Greece, Czechoslovakia, Italy, Yugoslavia, Rumania, and Holland.[32] The Hamerstroms thought those totals "undoubtedly conservative," but they knew that the effects were widespread and enduring. Lives had been saved; replacing some lost books, journals, and reprints had built collegial relationships. Reciprocal visits between the Hamerstroms and friends—abroad and in this country—continued for years. They exchanged visits and children with the Kramers and with Konrad and Gretl Lorenz, among others The bread they cast upon the European waters was to return, many times over, and for many years.

# 9

# The Return

## Deer and a Decision

*A man should be upright, not be kept upright.*

Marcus Aurelius

What they found at the George Reserve was far from satisfactory. In his first year (1941) as manager of the 1,260-acre biological study site for faculty and students from the University of Michigan, Hammy had reduced the deer herd from over two hundred to about 50 head. In his absence, numbers had again increased; he had to start over. He was required to preserve the maximum amount of salable venison, so he always aimed for the neck. Elva recalled, "We ate an awful lot of deer neck!"

The deer were one problem. "Improvements" were another. During the war, the Reserve's donor had dredged a channel in the southwest swamp, deepened and connected two small ponds, and built an airfield and a racetrack—changes that contradicted the Reserve's research policy. Hammy quietly asked Aldo Leopold, the trusted friend, to keep an eye out for some other situation for the two of them.

Congenial colleagues, like George Miksch Sutton, distinguished ornithologist and author, and Josselyn Van Tyne, an editor and ornithologist now honored by having the University of Michigan library named for him, welcomed the Hamerstroms back into a compatible circle. Working with Van Tyne honed Hammy's talents. He wrote and published four papers in 1947 and 1948. One, in the *American Midland Naturalist,* based on his ongoing collection of herbarium specimens, contained the factual apology—overbrowsing deer had limited his collection.

Fran spoke rarely about those Michigan years. She was able to contribute to the skeletal and bone collection of the museum there; she satisfied her curiosity about edible mushrooms through Alexander Smith, the noted mycologist. When she asked him to teach her those safe to eat, he handed her a pamphlet. "Learn to key them out," he said. She used that tattered, yellowing pamphlet yearly in Plainfield, when she gathered, dried, and stored the delicious honey caps that filled the woods some years, and also when people brought her specimens to identify. She was scrupulously careful.[1]

They visited Leopold at the time he was going through the process of submitting his manuscript of what was to become *A Sand County Almanac* to publishers. Fran was in his office shortly after one of a number of rejections came. A reader for Knopf had called the philosophical parts of the book "fatuous." Leopold handed her a page or two of the book. "Fran," he asked, genuinely puzzled, "is this fatuous?"

The Hamerstroms, along with Peg Hickey, Les Vogt, and H. Albert Hochbaum were in the group who worked with Leopold's son, Luna, to prepare the manuscript for submission after Leopold's death in April 1948. A three-page worksheet with editorial notes on *A Sand County Almanac* survives—a record of the collaborative Hamerstrom approach to editing. The first page is written on what appears to be an early title page, "Great Possessions." Here they jotted lists of page numbers and reminders: "Get from Pat data on chickadees for 62590," "Write Bill Schorger re the location of the monument to pigeons," "ant eggs p. 4," and the like; plus notes on spelling and capitalization conventions. A second page continues with longer comments. Hammy wrote about Leopold's musings on the mammoth, "Is there any reason to conclude that man finished off the mammoth? As a matter of fact, I object to these 3 sentences—they read beautifully, but they won't stand close inspection. . . . I am simply registering an objection at this point—if Aldo wanted it that way, it should stand." The third page reads: "I would have argued with Aldo over several points. I feel sure that he would have changed some of them—but I don't know which ones, and that is the crux of the whole matter.

Therefore I think that we should make corrections in . . . essentially mechanical changes—corrections of fact (if any are needed), and perhaps occasional deletions, but that we should not try to improve what Aldo has written . . . without specific direction left by him. This is a beautiful thing; better to leave a few things that might perhaps have been said differently than to risk taking liberties (no matter how well intended) with Aldo's own way of expressing things." What may be the draft of the letter they finally sent is in Hammy's handwriting but is signed "Hammy and Fran." The carefully qualified language is his, too. The notes, however, indicate that Fran raised some of the questions and prepared the list of possible emendations.[2]

This visit was before the Hamerstroms' return to Terrell's camp for the spring booming where Bob McCabe and Fred Greeley joined them. On 21 April 1948, McCabe drove down to Madison to check in at the office while Aldo was at the shack. There he heard the terrible news: Leopold had died, helping a neighbor fight a grass fire. Stunned, he drove back to tell them that the funeral was to be in Iowa.

It was the very peak of the booming season. Hammy, sure that the professor would have wanted them to continue with the crucial counts, did so. "I wish everyone could have heard the briefing Hammy gave that night," Fran recalled, "He poured all his emotion into telling the group everything he knew about prairie chickens." Sobered, conscious of the shortness of time, they went back to work.

The loss of Leopold was grievous. No one else had his influence, his lyric reach into public consciousness, his combination of warmth and clear purpose. His principles guided many of later Hamerstrom decisions: his sayings became their maxims, their practice was based on his example, and their beliefs often mirrored his. Hammy inveighed against economists, as Leopold had done. Fran quoted him when charged with neglect of their homestead: "Leopold said not to put too much time into maintenance." Years later, Hammy, in a rare show of feeling, told one of their gabboons that his last talk with Aldo had been shortly before his death. "We made plans to go hunting together the coming fall."[3]

Leopold may have influenced the Wisconsin Conservation Department[4] to hire the couple. Whether he did or not, by the next August Hammy announced to the Kramers that he planned to leave the

Reserve. He hoped their German friends would visit before any move.[5] He resigned in December.

I plan to leave the Reserve. I should like to stay through the winter, to finish several manuscripts. Please accept my resignation as of about the fifteenth of June, if that date is satisfactory to you.

Two things impel me to go. First, the Reserve continues to be over-browsed to such an extent that it loses a great deal of its potential value as a place in which to study vertebrate ecology. This is especially true in the case of the animals of the forest floor, the shrub understory, and the forest-field edges. To me, the loss is too great to be tolerable. Secondly, I cannot reconcile the University's avowed policy that the Reserve shall be kept as a "natural area," free of artificial disturbance, with the actual policy which permits such disturbances in exchange for an anticipated legacy. I thoroughly disapprove of what has already been permitted—changed water levels, dredging, the race track, the airport, tree planting, tree and brush cutting around the airport—and I see no real safeguard against further disturbance for the sake of expediency.

These criticisms are not directed at you or at the Museum's Executive Committee, for I realize that the contradiction in policy is imposed from above. I have never been sure just where it does originate. The fact that it is never out in the open is one of the especially distasteful aspects of the whole situation.

The issue is much larger than the mere presence or absence of artificial disturbance. The University certainly has the right to choose either policy. But I insist that it has not the right publicly to declare the one policy while privately it follows the other. I have raised objection to this situation before; my resignation is a final protest.[6]

He wrote the dean of the College of Literature, Science, and the Arts with a copy of this letter, hoping that he might influence the administration: "I do not offer . . . [it] in a spirit of bargaining: I shall leave no matter what concessions might now be made. . . . My reasons . . . might prove a useful straw if a determined effort is one day made to put the Reserve on an honorable footing. With that end in view, I have given my reasons as bluntly and baldly as I know how. It has not been an easy or a pleasant thing to do. I hope it will help the Reserve."[7]

His resignation brought him little but respect. One colleague declared that he wouldn't take the job for $10,000, another for any amount. A discreet academic acknowledged that there could be no research on small mammals as long as things remained as they were. A botanist blamed the deer for creating a botanically arid Reserve.[8] A letter Hammy wrote in 1954 to a member of the department of botany at Ann Arbor acknowledged the futility of resigning in protest. He felt by then that he had been naïve.[9] Hammy lived as he was: that he later called a position naïve does not indicate that he forsook it. A word like "honorable" and the strong reproof at the end of his letter of resignation define a man who truly lived his beliefs.

A search for jobs brought him offers to study prairie grouse in North Dakota, Montana, and Manitoba. No offer suited, for each answered the inquiry about employment for Fran with the news that a secretarial position could be arranged. As for living in Winnipeg? A city? Impossible.[10] Finally, what they had hoped for came: two positions for the fall of 1949 with the grouse project of the Wisconsin Conservation Commission—exactly suiting their experience, training, and their private dreams.

Kramers urged their American friends to come to Germany in the spring of 1949. Hammy could not think of it, but Fran wanted badly to go,[11] and soon he wrote excitedly to Gustav.

To come to Germany this spring is so improbable as to be almost ridiculous . . . I [first] . . . said it would of course be out of the question . . . There are at least a dozen good sensible reasons why we can't possibly do it. But we are going to unless something unforeseen turns up at the last minute. . . . I have been working furiously on manuscripts and the other "official" things that must be cleaned up.

Of the grouse, we are especially anxious to watch Lysurus and Tetrao. Tetrastes would be the one to miss, if we can't manage all three. . . . [Since] we are watching prairie chickens on the booming grounds, we generally go racing out from headquarters by automobile in order to get to the blinds before first light—and it is generally a race against time, with some miles to go and not quite time enough to do it in a properly dignified and impeccably scientific manner. That doesn't seem to fit our ideas of how to go about things in Europe.[12]

Happily, the university paid half of Hammy's salary during the Europe trip.[13] A joint passport issued on 8 February 1949 was valid for travel in the French and American zones of Germany. They wasted no time. After dropping the children off with Posey, they boarded the *De Grasse* on 5 March, heading for Freiberg and then Bavaria.

Conditions in Europe were difficult: civilian travelers were few and had no access to the PX; food, in short supply since the end of the war, was hard to find. The Hamerstroms were subject to military regulations; tight controls on changing currency attempted to limit the rampant black market. No matter, this grouse observation trip enlarged Hammy's expertise about grouse and deepened friendships seeded through the Action. Furthermore, they had time in their two-month stay to fulfill another, private purpose.

In addition to the ordinary requests for relief that came to them in 1947, they had been besieged with appeals from and about Adalbert Ebner, a Bavarian forester. On prewar visits to the United States and Canada he had been a guest in Leopold's home.[14] Leopold sent him up to the sand counties to visit the Hamerstroms in their graduate school years.[15] Ebner was something of a visionary on the subjects of forestry and wildlife education; he often talked with them until well into the morning hours.[16]

Ebner spent most of the war in Japan, working on a biota of Japanese plants at Sophia University.[17] When he returned to Germany in 1947 it was to prison, most of the time in Barracks 142/2 of the grim camp at Dachau, then a detention center for the American occupation forces, as a mistakenly reported member of the SS.

Ebner—frightened, embittered, and in poor health—sent frantic appeals: for food and medicine, references, affidavits, and possible connections or plans for release to many overseas contacts. The Hamerstroms responded with cheery letters and CARE packages, since his food ration was at the one-thousand-calorie level.

Neither Leopold nor Hammy could imagine this genial, philosophically inclined forester as "involved in anything actually or even technically criminal."[18] Ebner's family said that the American authorities had requested his arrest, but the Department of the Army did not respond to inquiries. Hammy found Max Rheinstein, a law professor at the University of Chicago,[19] who, after eighteen months in Germany with the Occupation forces, knew exactly which office in Munich they needed to address. Leopold wrote there not very long before his death; his letter, or his status, did the trick. A reply of 23 March 1948 announced Ebner's release twenty days after the receipt of the letter, after eleven months in prison.[20]

Ebner prepared a successful defense for the German denazification court. His institute had been abolished, and he could find no work. He asked friends abroad to find a position, and flooded the Hamerstroms with letters: long, single-spaced with a dim typewriter ribbon, or written in a spidery script; letters in English and German that ranged from the sentimental to the angry, from humble to overweening. Some were flattering—"Hammy, of course will bring his superior brain to the question"—or reproachful—"what nonsense she talked about politics!"—some unctuous, others grandiose. Reading them today is burdensome; to have answered effectively might have been overwhelming.

The Hamerstroms believed in him, and they had reason. Papers they found in Munich recorded his status as "unfit" for the draft in the town of Dachau. He belonged to no military group anywhere, including the SS.

But when Ebner finally got an offer from the University of the Andes in Merida, Venezuela, the necessary exit permit was denied by the American authorities. Hammy and Fran were outraged. They were convinced that Ebner was innocent of not only the SS charge but of the further allegations–that he had been a spy—upon which the permit refusal was based.

Birger Berg, an American forester in the Office of Military Government, Bavaria, believed that charge. Hammy talked to Berg: "Mr. Birger Berg (Chief of the Forestry Section O[ffice of] M[ilitary] G[overnment], Bavaria) told me outright when I was in Germany last spring that he would not allow Dr. Ebner to leave Germany. I was able to investigate Dr. Berg's charges before I left, and I am convinced they simply are not true. . . . I suspect that a personal grudge . . . is the basis for the difficulty . . . especially as the two men were in graduate school together at Ann Arbor."[21]

Hammy and Fran (whose German was invaluable) had found all they would need in Munich, then a struggling wreck of a city much damaged by aerial bombardment. They tracked down Ebner's court papers and files at the Ministry of Education and the office of the Military Government of Bavaria. They interviewed various officials at the university in Munich with which Ebner had been affiliated. A Scottish woman, a Ph.D. who had worked there, told them how hard it had been to get Ebner's permission to go overseas, not what one would

expect had he been on an official mission. German authorities had opposed Ebner's foreign travel and friendships, and his prewar interna-
tionalism had endangered his career. No shred of evidence supported
another charge—that he had been paid a thousand gold marks per
month. A letter to him identified that sum as a normal professorial
salary, but he was never a professor. Rather, he prepared a thousand-
page biotica, as the letter in early 1949 (from the Japanese forester
someone must have contacted) indicated: "I published forestry book
with him."[22]

Hammy's contact in Washington, Tom Gill in the forest service,
had previously cooperated in the Action. He advised working with the
appropriate Wisconsin congressman. Hammy did so and told his congressman of the injustice that had been done. Months later, a cable
from Congressman R. F. Murray to John J. McCloy (U.S. High Commissioner in Germany) got Ebner his permit to leave.[23]

It was well beyond the intent of this book or the possibilities of this
author to search either German or American archives further. Hammy,
a practiced researcher, brought thorough and persistent scrutiny to the
matter. He acted to fulfill his credo of living with respect for people as
well as for the natural world. And he and Fran continued to befriend
Ebner who, from Venezuela in 1950, continued to write to Plainfield
with essays (in Spanish), accounts of his bachelor hardships, pleas for
letters ("I am lonely"), and criticisms of the rote education system at
his university. He wanted to borrow teaching materials for the coming
semester. A four-page letter that took him six days to write and included a paean to the Hamerstrom marriage[24] implored their help in
getting an exit permit for his German fiancée. They did as he asked.
When, back in Bavaria, he fell ill in 1954, Fran and Hammy arranged
an emergency $100 donation through Kramer, who sent Adelbert Ger-
man marks in exchange for their providing an equivalent sum to a German student in America. They loaned Ebner money.

Ebner visited them in Plainfield, and in the late 1960s his step-
daughter, Silvia, came for an unsuccessful stay with the Hamerstroms.
Their unhappiness with her behavior did not harm their long-standing
friendship with her stepfather, as a 1965 letter from him attested. Alan
remembered his parents' puzzlement that their friend Ebner had not
repaid the loan they made him. "They just couldn't understand it."

The Hamerstroms based their actions on what they deemed was right; should others respond unworthily, they withheld judgment. Fran preferred to tell amusing and dramatic stories about that European trip. They went to a nobleman's party on the streetcar, she said, and changed on arrival into the required formal clothing they carried in a bag. Another favorite account was of a hike in the Dachauer Alps.

They were climbing with Adalbert Ebner and saw a snake. "Is it an adder?" asked Fran. She knew that the only poisonous snake in Europe, the *kreuzotter,* lived in that area. "Not at this altitude," said Ebner. He had no idea she would reach for it. It struck and bit her arm. Hammy stayed calm and applied a tourniquet.

"We'll trace the movement of the swelling as we go down." It was two hours to the nearest village; in his haste, Ebner fell and sprained his ankle. He limped along with them until Hammy inspected the arm. The swelling was moving up the arm, about an inch an hour.

"Leave me here," Ebner insisted. "I will slow you down. The track is clearly marked to the village, you'll find a doctor there." They went a long way, part of the time in the twilight, to a village and found a physician in his home.

"I've been bitten by an adder," reported Fran.

"Madam, I hardly think so." She displayed an enormously swollen arm. Alarmed, the physician demanded, "How do you feel?"

"Hungry!" she replied. They hadn't eaten for hours, so the doctor's wife provided some soup. He arranged for them to be taken to a nearby town where Fran was treated with antivenin. Hammy waited for a guide, at his host's insistence, before he went back to bring Ebner down. "They were right," according to Fran. "You can't negotiate those mountains in the dark." Hammy made light of the incident. To him, it was "Typically Fran!"[25]

The trip not only satisfied Fran's longings for sophisticated and exciting experiences but also displayed Hammy's qualities to German foresters and scientists. From that setting, at once exotic and instructive, they returned to the rustic sand counties. Hammy became the new leader of the Wisconsin Conservation Department's grouse project.

Adalbert Ebner had called their marriage "perfection." "I see that the so-called 'fun'—it is far more, it is the art of life—it is the 'way' in the Japanese sense of grouping a whole life around an idea. . . . You work together—your work in common—your love is in common—you have fun together . . . for me you are the perfection."[26]

Perfect or not, they were fond, flexible, resilient, and always intimate. There was no question about the almost palpable tie between them. Roger Tory Peterson spoke, in his introduction to Fran's book *Harrier, Hawk of the Marshes,* of her "adored" husband. Determined to live fully the life together for which they had prepared, they faced the question Leopold had asked with forceful simplicity, well back in the 1930s: "Is the Prairie Chicken Hopeless? . . . In every state of the north central region except Wisconsin, the restoration of the prairie chicken is regarded as a 'lost cause.'" They trusted his answer: "The trend of the evidence . . . is to the contrary. It indicates that chickens respond even more readily than other cyclic species to management measures."[27] Their task would be to design those measures.

# 10

# The Setting, the Task

*What is man without the beasts? If all the beasts were gone, man would soon die from a great loneliness of the spirit. For whatever happens to the beasts soon happens to man.*

—Chief Seattle

As head of the WCD's Grouse Project in 1949, Hammy undertook administrative and supervisory responsibilities managing ruffed grouse and sharp-tail, members of the grouse family elsewhere in the state. The major emphasis, however, was on prairie chickens—especially in Portage County, home of the great Buena Vista marsh.

Driving the well-known highways, he stopped at the spot where, in the spring of 1938, he had seen odd shiny streaks in a pattern of interrupted stripes stretching for rods in a field near County Trunk O. He had walked out, knelt, and felt the firm strips and marveled at the compacted scars of the plowshare that stayed put after six inches of fertile topsoil blew away.[1] Surprisingly, the dismal landscape, marked by gnarled oaks and scraggly Scotch pine, still appeared much the same as it had then. Gray unpainted farmhouses still stood on county roads that wound past hillocky fencerows.[2] Some were deserted; others connected to electric wires, for the nine-year-old Waushara County Rural Electric Cooperative had brought transforming electricity to many farms. Solid brick community buildings built by the WPA added substance to the main streets of small towns; yellow buses made entrepreneurs of the residents who drove children from widely spaced farms to consolidated schools.

The population was smaller than in 1939; the wartime defense boom in the cities had made life on a sand farm comparatively unattractive. In 1934 the editor of the *Necedah Republican* could logically defend the right of the poor man to move into submarginal land—the traditional "haven, where he could retreat and live, with little initial investment."[3] Now what Aldo Leopold called "the hammer of development" was beating on the "anvil of wilderness."[4] The light soils, wetlands, and oak savannas were yielding to new agricultural practices. To the west, large, profitable cranberry operations transformed townships that had originally been covered in "water from six to forty inches deep, with marshes destitute of timber." Tamarack swamps "unfit for cultivation . . . [with] no permanent settlers,"[5] had become fast-growing pine plantations to supply the paper mills on the Wisconsin River. The "great dead heart" of Wisconsin was beating once more.

In Waushara County, a citizen planning committee labored for eight months and decided: some fifty thousand acres of unsuitable land should be retired; its swamps turned into game refuges, hunting areas, or marsh hay fields. (Had their assessment held, the Hamerstroms would have had a simpler task.) Merely marginal land, eighty thousand acres of it, could become 360-acre farms with shelterbelts, woodlots, and feed patches for wildlife, that would produce a healthy livelihood for families—for $25 per acre. The committee envisioned bountiful gardens and pigs, cows, and chickens—the previous era's agrarian dream. They ventured a cautious prediction: "Approximately half of the land in this area lies within fifteen feet of the water table. . . . research . . . [may] develop a practical irrigation system for many of the farms . . . completely [revising] . . . long term plans."[6]

The report was outdated almost before it was read. Irrigation was about to transform farming. Long time residents, intrigued by Professor A. R. Albert's demonstrations at the Hancock Experiment Station, bulldozed pits from which they pumped water through labor-intensive hand-moved pipes to fields of green bush beans or matted cucumber vines where once only moss and sandburs grew. Enterprising farmers began to move into the country. One of them, Bob Moldenhauer, brought along his 1951 master's degree in Soils from the University of Wisconsin. He and fellow graduate student Dick Corey devised a system more efficient than a pit. Bob (with a 1941 Chevy) and his partner (with $400 in the bank) borrowed $5,000 and persuaded a Coloma banker to sell them a 320-acre farm. Bob's bride brought a dowry that put a down payment on an old Ford tractor. The dealer looked them over. "Pay me the rest when you get some cash," he said.

They bulldozed a long, shallow trench into which they pumped and forced powerful, spaced jets of water, thus putting down five twenty-foot wells. These, linked to a centrifugal pump, drew four hundred gallons of water a minute from underground old Lake Wisconsin. Thus, they could irrigate forty acres. Corey, inducted into the army in 1952 right after he got his Ph.D., sent what he could from his stipend to support his investment. Eventually they irrigated 240 acres of beans, sweet corn, and cucumbers.[7] They pioneered, just as the Hamerstroms were doing. They made the old farmhouse habitable. They walked six miles into Hancock in a 1951 winter snow when they could not get the car out. They worked all night, carrying and laying out pipe. When they found no market for the fine tomatoes they grew, they tightened their belts and tried something else. They endured the spring winds that blew out their carrot seedlings. They earned the neighbors' admiration: "He was a worker, all right. And he did it before anyone else thought of it." Moldenhauer's early death in the late 1950s created folk-hero status with his many admirers.

Fran and Hammy, in contrast, branded by breeding and purpose as the ultimate outsiders, arrived just as mechanized farming began its takeover of the area. Their diligent efforts were counter to the existing culture, their focus unprecedented. Their search for ways to increase prairie chicken populations did engage some residents, but it baffled others and enraged those who deemed that land was for production. When I asked Jim Hale, the director before Hammy of the grouse project and later head of the department's Wildlife Research section, if anyone besides the Hamerstroms could have saved the species, he looked startled and then replied, "Oh, I think so. Some other scientist would have found a way—to be sure, a different way. It would have happened."

Perhaps. In the sand counties, however, just emerging from hard times, many residents had to focus on improving their own livelihood. The combination of devoted effort, meticulous research, ingenuity, tenacity, and the personal magnetism that explain the Hamerstroms' success, would have been difficult to match. Who else would have entertained, educated, fed, and housed the seven thousand boomers that viewed the prairie chickens from their home? A more conventional approach might well have been too slow. Yet, paradoxically, the very qualities that enabled them to succeed added to the reputation that made them vulnerable to censure.

We can only imagine presettlement Wisconsin today, for the state then had some two million acres of prairie. The inevitable shrinking of those expanses caused the threat of extinction under which the Hamerstroms worked. Leopold had described the northward movement and decline of the Wisconsin prairie chicken range in his 1931 *Game Survey of the North Central States.*[8] In Dane County, he saw a big flock booming and nesting in 1932, yet two years later, with little apparent change in habitat, there were only fourteen birds, five in 1935, and none in 1936. What was to blame? In 1931, he estimated that some four thousand birds remained in Portage County, where the bulk of the Buena Vista marsh lay.[9] Even there, Wallace Grange predicted, the prairie chicken would be too few to hunt in twenty or thirty years. He was close: the season in Wisconsin closed in 1959.[10] Grange, at the time, had not experienced the land boom in the sand counties that required the Hamerstroms to revise their conservation strategies repeatedly.

In 1950, their immediate need was for data on population, mortality, flock composition, local shifts, and sex-and-age comparisons. The information had to be consistent over large areas and significant time periods. That meant regular yearly winter trapping and banding to provide identifiable birds for the count at spring booming. This task was so vital, and so demanding, that the Hamerstroms' chief, Cy Kabat, a diligent conservationist, assigned himself and his entire wildlife research staff to serve as observers at the peak of the six-week season.[11] All this, plus what they learned about chicken habitat would go into their management plans. Should the data relate to the cycle, they might be able to understand why prairie chickens had disappeared over most of the state.[12]

Fran was exultant. She wrote to the Kramers, "We have a gorgeous view of fields and woods of pine and oak. To the North lies our main study area of sixty-four thousand acres and to the south . . . our old study area of fifty thousand. . . . We are forced to concentrate part of

our efforts on . . . land use, human ecology. . . . Hammy has to spend
. . . time on reports 4 times a year . . . about thirty-five people read
them . . . it is not sensible to write long excellent reports for so few
people to read."[13]

Her complaint about reports went on through their years with the
DNR; Hammy quietly complied with painstaking records of their findings. Seeing the complexity of the task ahead of them, he reflected. "It
won't be easy without Aldo." And then, brightening, "Maybe we can
find Os." They did, down in a hole, digging on the Petenwell hydro-
electric dam. Os Mattson reported, "A shadow darkened the sky. I
looked up; there they were. She was in heels. So, of course I went back
with them, though I took a cut in pay."[14]

The Ware place was now past mending, and camping at Terrell's
while Os went scouting would have to do. Soon Os found an ideal set-
up; he drove excitedly to their tent to describe it. "It's 240 acres of
wooded land, with marshes, old corn fields, and a house that goes back
to Civil War days. Wait till you see it!"

They saw a huge house, groves of noble white pines, a sturdy barn,
a windmill, a roofed vegetable storage cellar, and a shed for the car. It
was appealing: when Hammy saw wood scraps in the big room behind
the entry he spoke as if they were owners: "We'll stack wood outside;
trapping gear belongs here."

They admired the wide transoms, good proportions, porches on
the east and north—but almost every window was broken. "Glass is
easily replaced," said Hammy. As they moved from one high-ceilinged
room to the next, strips of hanging ceiling wallpaper whispered in the
wind. Bare lath gleamed through gaps in the plaster, a stack of doors
stood against the studs of the unfinished attic upstairs. Fran rubbed
a foot along its dusty floor. "Hammy! This is oak flooring!" she exclaimed.

"Os said they held dances up here! Perhaps they did."

"Think of it, Hammy! We could bunk crews in this room."

They laid out a possible apartment for Mary and Os, with a kitchen
and a private entrance. A small front room by the front door would
hold their piano and Hammy's record collection: "Our music room!"
The spacious corner chamber, fitted with a wall of bookshelves,
matched Hammy's ideal of a study.

The Plainfield home in the late 1950s. The lawns, manicured by Hammy, the stately elms, and the woodpiles ready for the coming winter bespeak their pleasure and pride in their home.

Suddenly, Hammy paused and cautioned, "Don't set your heart on it, Fran. I'll check the roof and foundation. We mustn't take on too much maintenance." The house proved sound.

They took eight-year-old Alan and six-year-old Elva to Chicago to be met by Posey, who took them back to the family home in Milton, where she was now living alone. They left Terrell's camp, pitched their tent in their new dooryard, and set to work: replacing windows, hanging doors, tearing down wallpaper, patching and painting walls. Electricians wired the house; the local truck firm moved their stored furniture from Michigan. This was to be their base, their home, and their hotel for the rest of their lives. "We all envied him his field station," commented Hammy's long-time colleague Jim Hale.

Meanwhile they had to build needed traps and gear. Hammy designed and made two scale boxes of waterproof plywood: the scale sat in the bottom, rather as a sewing machine might, and the deep lid made a seat when set on the snow. He built two wooden tool kits about the size of a large briefcase, one for each of them for needed supplies:

bands, pliers, wire, record sheets, tags, and first aid material for any injured bird.

They accomplished enough to allow a respite. Posey agreed to enroll the children in the Milton schools so that Fran and Hammy could
make a September swing through northern Douglas County.[15] They
hunted a little in the flourishing sharp-tail area of the burned-over
northern counties and then circled back through North Dakota to acquaint themselves with a wider range and different habitat.

Hammy repeated a proven grouse project mail-in technique of collecting the wings and tails of birds bagged by hunters. He distributed
fifteen thousand reply envelopes that fall; the returns helped him make
estimates of the existing population.[16] The second year they had a list
of 2,468 sportsmen, dubbed "the wing-tailers," and distributed collecting envelopes all over the state. This practice continued until hunting seasons on chickens closed. Hammy expanded an established food-
patch program with sportsman Leslie Woerpel, who was later to bring
twenty-nine organizations together in the Wisconsin Federation of Conservation Clubs, Leo Gwidt, Stevens Point druggist, and Vilas Waterman, a small farmer, who manned nine feeder stations and urged
neighbors to plant winter food patches.[17]

Soon the yearly call to help register deer shot by hunters in November, the "deer season madness," as Hammy called it, was upon them.
Then came an all-out effort: the winter trapping. Seven mornings a
week they donned layers of clothing, wool socks under heavier wool
stockings, parkas, mittens, and scarves. They laced their moccasins
or boot packs, grabbed shovels and axes and a bunch of old socks in
which to stow the birds for weighing, a burlap bag stuffed with extra
bags, a tool kit—all the essential gear. Fran stowed snowshoes and
equipment in her battered telephone lineman's truck; Hammy loaded
the station wagon, and they took off in different directions.

When roads were blocked with snow—how it blew on the marsh!
—they strapped on snowshoes and trekked in with a load of bait at
each trapping station. No birds? Back to the welcome warmth of the
car for the run to the next station. If they found birds, they handled
them gently, and always with bare hands, whatever the weather. Gloves
would pull the drumstick feathers out, leaving the birds vulnerable
to freezing. They eased each frightened chicken into a sock, where

it would lie quietly in a bag—with no more than four others—for processing before release. Weighing, aging, sexing, banding, and a quick
assessment of each bird's condition took only a few minutes. Then it
was on to the next trapping station. It was slow work, and it could be
disheartening—day after day with no catch. Ten birds per trapline was
a good day. Then, suddenly, came a monster catch.

John Rozner lived on a farm at the north edge of the marsh some
twenty miles away. He had complained that he couldn't get the corn in.
"It's still shocked out in the field." Here prairie chickens congregated.
They ate from the outside of the shocks. More birds came. On 12
March, John called the Hamerstroms.

"Say, there's lots of chickens in my corn. The corn's almost gone,
but the birds still come every day."

"Let's go." Hammy sounded excited. "We'll need all the traps."
Forty-eight clumsy traps, each five feet long, were scattered over several thousand acres of marsh. As Os Mattson and Hammy brought
them into the barnyard in batches and loaded them, Fran drove them,
four loads of them, to the Rozner's place. The truck held four traps,
stacked upright in the bed, and it hauled a two-wheel trailer, loaded
with eight more bulky traps.

Four of them—Mary Mattson helped—put all those traps in Rozner's fields and baited them before the chicken pack came in to feed.
The morning catch delighted them; the evening one almost over-
whelmed them. They processed birds until well after dark.

That winter's trapping gave them three hundred chickens and
seven sharp-tails, a record that brought a quandary: How could they
observe all the booming grounds where the banded birds would soon
be seen? Fran typed and retyped a letter with ten carbons and sent it to
"anyone I could think of who might come."[18]

That letter drew over one hundred volunteers, still too few to cover
all the booming grounds for the entire six-week season. Evening after
evening they came, heard Hammy's briefing, bedded down in the ballroom, and rose at 4:00 A.M. to be served a hearty breakfast before dis-
tribution among the blinds. They saw chickens booming and sometimes
copulating, read band numbers, took notes, and drew schematic maps
of the booming ground they had observed. Many of them asked to come
back. The long process of establishing booming as a not-to-be-missed

Hammy in the early years of the prairie chicken project, releasing prairie chickens after weighing, banding, and processing.

experience with the Hamerstroms as hosts and authorities had begun, setting a pattern that continued for more than twenty years.

They located booming grounds by revisiting those they had known in the early 1940s and by talking to residents. Carl Hakes told them about the bluff where he had hunted as a boy about twenty-five years ago. (The Hakes's prairie, acquired when the project began to buy land, would become a prime site for observation and experimentation.) With such approaches and careful scouting, they found forty booming grounds in the twenty miles between their home and Stevens Point.

Hammy's explicit instructions to area mangers in other areas of the state describe the means of finding flocks. They must start at "the first gray of dawn" and " be discontinued about 1 1/2 hours after sunrise. Use clear, calm, cool (below 50°) mornings for finding grounds. . . . Stop every half mile, shut off the motor and move away from the car to get clear of the noises of the cooling engine. Listen for at least one minute. When sharp-tails or chickens are heard, triangulate on the sound from at least two stops. . . . Unless birds can be seen from the road, make sure that the ground has been . . . accurately located . . . before attempting to walk in on it: chickens which sound as though they were surely in the next 40 may prove to be a mile or more away. Cover the transect thoroughly at least twice. . . . Run transects in the opposite direction on the second search."[19]

They recorded the results of such searches, but since maps from earlier years were out of date, they began mapping anew. Accurate maps would always be essential; they reconstructed theirs regularly, using recent aerial photographs, farmer reports, and fieldwork. They walked every field with tracings of U.S. Geological Survey maps, marking trails, roads, ditches, dams, and dwellings. Hammy keyed the vegetation by species: at least ten types of hard- and softwoods differentiated by size and stand—solitary, windbreak, plantation, or scattered. He noted upland brush (sweet fern and sumac); lowland brush (tag alder, spirea, willow); grasses, forbs (such as goldenrod), and mixtures; cropland and marsh, sedges and marsh grasses. The key filled two pages with his fine backhand and color-coded, carefully drawn symbols.

He and two men of the forest inventory spent hours on the task; Os added cultivated fields and herbaceous types the next summer. All this and the results of a winter food survey went on one acetate per section. Finally, a skillful conservation department engineer combined the acetates into a comprehensive map. "A considerable job," commented Hammy, "as the air photos and the acetates traced from them were not always to scale."[20] No wonder the mapping took a full two years to finish.[21]

After all that preparation and a fine trapping season, observations the first spring were disappointing. Wind, storm and clouds prevented their finding all the color-banded chickens they had trapped that first winter, and the ink on some bands was unreadable. Hammy wrote to a friend, "Maybe there's a good future in selling bonds, or tatting doilies. Pooey."[22]

They simply went on, applying disciplined curiosity to everything they could discover: food habits, nesting, population growth and decline, sex and age ratios, seasonal movements, and the effects of hunting and disease on Buena Vista grouse and in nearby populations in Portage and Adams Counties. "To find out what the birds ate . . . we had to get the crops and stomachs. Checking for parasites was important. During the season, we'd go into bars, I on one side of the street, a reluctant Hammy on the other. We offered to pluck the birds hunters were bragging about. Most of them said yes. I'd get their car keys . . . take the birds to the back of the bar and pluck them into a trash barrel. They'd get the exquisitely plucked bird, absolutely table ready, and I'd get what I wanted—the crop and gizzard."

They called it the "Hamerstrom Plucking Service." She thought it "great fun," but Hammy detested it—boozy patrons pushed confidences through his polite reserve. They sent dead birds and blood smears from live ones to the pathologist in Madison who found roundworms, gizzard worms, tapeworms, crop worms, intestinal worms, even grouse eyeworms. Minor setbacks—as when the jars, full of crops, gizzards, and intestines for the pathology studies exploded—interrupted the routine. They had to pick up the material. "Can you imagine the stench? And the two of us, picking up the specimens in the dustpan? We had to save the specimens."

Improved bands were a priority. Hammy wrote Elmer J. Haas of the National Band and Tag Company of Newport, Kentucky: "The color bands make it possible for us to recognize individual birds if we can see both the color and the number . . . if we can see at least the color or . . . colors without reading the numbers, we know where the bird was banded even if we cannot then tell which individual it is." Black numbers on white or yellow bands were best; an ideal band would be wider than existing models, with a durable white ink on a dark band with a secure fastening. Lost bands meant loss of the precious data, hard to accept after the rigors of trapping the shy creatures.[23]

Aging the chicks could help determine the age of trapped chickens. Summer was the time to find chicks in the wild, but finding them was "devilish hard work." One of their first helpers had tagged muskrats. "Why not put a tiny wing tag on newly hatched chicks? Some hunter will bag one of those birds in the fall; when they turn it in, we'll have the age to the day."[24] Hammy offered a $2 reward for each wing-tagged bird turned in, but few such finds resulted. How could they then find live broods?

"Let's go see the strippers!" Fran suggested. These local men who loaded the sacks of seed onto the trucks following the bluegrass harvest saw chicks. Soon the crews were proudly reporting numbers. Fran and Hammy often accompanied them, running down the older chicks and simply picking up the very young ones. Most of July the two of them and Os followed one crew or another through the long summer days.

Soon Hammy was able to compare maps and records from the 1940s with current data. He saw that most of Waushara County was now second- or third-rate, with only a few active booming grounds at the peak of the cycle. "About the best we can do there is to be sure that there is winter food and to keep what grass there is." The Adams County range was precarious: two growers, owning almost five thousand acres on Leola marsh, had plowed several eighty-acre tracts of grassland. Portage County's Buena Vista marsh was the spot for management. Agronomists maintained that agriculture there was not likely to intensify; the drainage district was inactive; the bluegrass seed industry compatible. Those grassy acres, plus the edges and meadows on scattered ranches and dairy farms, would provide the prairie chickens with the wide horizons necessary to their survival. Here was the place where long-term research could proceed and they could finally put down personal and professional roots.

And Fran and Hammy had the house for it. They hated renting, even for one day's salary a month. They had put too much effort and money into their home to think of losing it.

Boyd Walker, our landlord, was about sixty and worked in Milwaukee. . . .
He married his landlady, Goldie. He said we could have the whole place for $9,500 and would get first choice when he decided to sell. Still, people kept knocking on the back door, saying they had been sent to see it. That made us uneasy. They would ask, "Is the house hard to heat?"
"Not as a rule," we'd reply. "The stoves need stoking fairly often, but when the east wind blows, snow comes in the front door." It was perfectly true; it did. We'd tell the children to take them upstairs to see the view from the front windows. One of the more conspicuous things they could see was a large hole in the roof of the porch. One day, as Boyd told us, Goldie reminded him that he only had one lung. "I want your estate settled. Sell that place!" she said. He obeyed; and didn't change the price. It took just about all we had in the world, but we paid cash. How happy we were!

And they were, except that Hammy had too much to do. Running the grouse program statewide; maintaining time records; reviewing reports; handling hiring and budget; and negotiating on matters as small as a hopeful breeder's request for a live-trapped bird with the authorities in Madison, cut his time in the field. Fat quarterly reports recorded the copious details of investigations. And, ninety miles from a good library, he sent off regularly for material on history, geology, and forestry, and for reprints from colleagues that would keep him up to date.

He spoke to groups, especially those of hunters. He answered each letter and memo, typing in a fast three-fingered fashion on his sturdy Underwood manual most evenings. A farmer asked about winter feeding on a "hunting forty." Hammy replied at once, "It seems that 100 years ago chickens may well have been migratory. If wintering conditions were bad in one place, they could find another. Now, with so little range left, we've got to take care of each flock right where it lives."[25]

It became clear, however, that even with Fran—innovative, energetic, invaluable—the job was too big. Only Os and Mary Mattson's dedication enabled them to handle the two Buena Vista traplines that covered almost thirty miles each. Luckily, people wanted to work with them. Dick Hunt, a recent graduate from Madison, had joined them that first full work-summer of 1950. A skilled trapper, he handled the vital nesting and brood survey while they went to Europe. Fran left him to care for her pet fox. "She kept it in the house," he recalled. "It escaped. Maybe," he said with a sly glance, "I left a window open." He and Os built fifty traps. Come winter, Dick Hunt moved north to live with a farmer and handle the trapline up there.

From then on, there were always helpers about, learning under an informal but highly selective system. Some were one-time aides; others returned several times. Strictly speaking, only those who worked with her on hawks and owls were designated as gabboons, but many helpers appropriated the title.[26] Fran and Hammy delegated daily routines to these helpers. They used the time saved for projects like surveying the Little Eau Pleine area, a marshy river valley to the west, rich with wildlife that became the Mead Wildlife Area.[27]

"I learned to be thorough from Fred. Few biologists I've known were as particular as he. I don't know how many would have stood the grind of the booming season—three months, essentially, out of every year," said Hunt. "Once we ran the trapline at 46° below. I'll never forget it. Fran and I chugged out in that old Model A truck. It was so cold I could trace her movement down the other end of the line by the plume of her breath. You had to admire the way she stuck to it." Hunt learned to call a spade a spade. "Fred was about the most honest man I ever knew. And he was the boss. 'Isn't that right Hammy,' she'd say."

He learned biologist etiquette. "Once I found a bird killed in the trap. I was green; I took the pinnae off the carcass and put it in my hat. When Fred saw it he was angry. "'Where'd you get the feathers?'

"I told him. His voice was quiet, but his eyes flashed. 'What did you do with the bird?'

"I threw it in the bush."

'I want you to get right back up there, find that bird, and bring it in. Then you skin it and prepare a study skin. You can sew the pinnae back on.'"

Dick explained. "To make a skin is exacting. You cut the bird down the chest, and gradually work it out, knees first. Cleaning the wings is tricky, but the hardest part is to cut around the eyes without damaging the eyelids. You take out the brain and make sure all flesh is off the bones and the skin. On that bird, I had to sew up the damaged back before I could treat it with arsenic and borax and then stuff it.

"We hunted together one time on the Big Eau Pleine. Fran saw a grouse so far away I would have let her shoot at me at that distance. She aimed—shot—and she hit it!"

"Fran shoots," someone said, "and God strikes it dead." Dick laughed, then became serious. "But I learned from him to enjoy the chase. It really didn't matter if you got anything or not. And you know *he* was the one that got me into grad school."

They, not their work, generated immediate interest. Fran laughed as she described an inaccurate *Milwaukee Journal* headline: "Hagerstroms Quit City Life to Study Wild Life in Woods."[28] The two tons of books and documents that they brought with them, her debutante background, and the fact that she did not know who Wallis Simpson was got more space than the need for food patches.[29] "Never mind," said Hammy, "people will begin to learn that prairie chickens need attention."

Shared work, learning, adventure, overcoming difficulties, and becoming known—it was the Hamerstroms' dream. Underlying it all was the land. They could go crane watching with Hazel and Wallace Grange and actually find the gray, gawky, shy birds. They could savor a sunrise on the marsh, or delight in the harriers' sky dance. The children and Fran took magical moonlight walks.

Their circle of local friends, supporters, and helpers was growing, among them conservationists Wallace Grange and Les Woerpel, Plainfield postmaster Harry Walker and his brother Pete, local residents Boyd Kelley, Shirley Barnes, and Tip Booth of Hancock; Plainfield farmers Lloyd Conover and Carl Hakes. They were happy, purposive, positive. They had high hopes.

# 11

# Booming Chickens and a Land Boom

*But in science the credit goes to the man who convinces the world, not to the man to whom the idea first occurs.*
—Sir Francis Darwin (First Calton Lecture, the Eugenics Society, 1914)

It seemed they had everything—commitment from the WCD, two salaries, an ideal home and field station, and a research-ripe situation with prairie chickens before them. Fran, promoted to Conservation Aide on a 60 percent time appointment, was paid almost $200 a month, with the bonus of health insurance for a deduction of $4.80 on each paycheck. They had made friends in the community; their superior, Cy Kabat, was warmly engaged. Above him the organization grew in complexity, with an appointed conservation commission and commissioner to determine policy—a structure quite subject to political pressures. All seemed hopeful.

In 1952, when an expected drop in the spring count showed only about half of the cocks of previous years, Hammy wasn't too concerned. He worried about the sharp-tail numbers in the Namekagon Barrens. There, in burned-over Douglas County, young oaks, jack pine, and abundant blueberry patches provided good habitat. Juvenile birds grew fat on pin cherry, birch, and aspen buds; grass and low vegetation provided cover for nesting and rearing the young. An occasional fire preserved this natural balance: hunters enjoyed good hunting, their wives, plentiful berry picking.

Then came "progress"—fire suppression and Smokey the Bear—causing the inexorable return of land to forest. When an uninformed legislature passed a forest crop law that rewarded planting trees, the heart of the Namekagon Barrens was planted to pine that summer. Pine plantations are deadly for sharp-tails; this bird population, already in danger, would decline further, perhaps to extinction. Ernie Swift, the department chief, charged Hammy with making an unaware public see the problem.

"It is up to you to sell the sharp-tail dilemma," he declared. "We can't wait for that booklet you are working on; we don't have that much time." They had not been hired to manage public relations, but they scheduled more speeches and rigged up a flannel board to dramatize the plight of the birds. Fran made the board from an old conservation department sign, painting it in broad horizontal bands of color: blue with white cloud shapes at the top, sandy yellow and mauve earth tones on the bottom. She over-painted stained white flannel with a palette of black, soft greens, browns, and orange. From it she cut images of trees, shrubs, a hunter with a gun, a flock of birds, and a single cock sharp-tail. She stapled tufts of tan grass on a long strip of tawny flannel and patted it into place in the lower center. On this background of barren land she built a sequence in which her figures revealed the eventual result of the growth of vegetation—a cock, dancing alone. It impressed audiences. Hammy saw heads nod to his forceful assertion, "A pine plantation is a biological desert!"

Fran wasn't satisfied. "It's no good talking to twelve people here and twenty there. We'll just have to go on television."

"Television!" he expostulated. "Fran, we don't know anything about television! We've never even watched it!" She pulled a small government bulletin from her purse.

"Here! I paid ten cents for it!" He took it, examined the title: *How to Appear on Television,* he read. He handed it back. "What does it say?"

"I mustn't wear distracting patterns; you should wear a light blue shirt. I'll go down to Goult's Furniture and Funeral Parlor tomorrow to watch a program or two." Soon they were on WTMJ in Milwaukee, the largest TV station in the state. At the end of their program Hammy smiled. "I think Aldo would have approved," he said, and added, "We don't get to the big city often. Let's go dancing!"

Then Gordon MacQuarrie, perhaps Wisconsin's most popular outdoor reporter, came through for the sharp-tails: "The Hamerstroms have diagnosed the case . . . and written the prescription. . . . Is Wisconsin going to do something about it, or are we to go through that particular kind of public agony that marked the end of that last heath hen on Martha's Vineyard? Are we going to lie in bed and refuse to go down and see who is knocking at the door? And then when it's all over are we going to erect a lovely monument . . . as we did for the passenger pigeon at Wyalusing?"[1]

The president of the Citizens Natural Resources Association of Wisconsin, Jess Walker, led a campaign against further tree planting in the barrens at the Conservation Commission. Fran recalled that he went to the hearings in spite of being ill. She went on: "It was our first really effective campaign. He found the commissioners attentive; they even sent a telegram up to Solon Springs that very day, ordering planting stopped until they could make a final decision at their next meeting. Walker was snowed under with calls, letters, and telegrams that did the trick." They were learning about political action.

Three thousand copies of their booklet, *Sharptails into the Shadows*, appeared in 1953. It won immediate praise. Walter Scott, the remarkably well-informed game manager who had become game commissioner Ernie Swift's assistant, called it a "poetic masterpiece."[2] He wanted more copies. "How can three thousand be called 'popular?' It hurts to put this under lock and key." [3] A Michigan game division administrator requested three hundred copies with a query: "We have an administrative curiosity as to the extent to which this was written by the technical people involved and to what extent the editor may have reworked it. One of our problems is combining a good job of writing with a sound technical study. You appear to have solved this problem in this bulletin."[4]

Their reputation in professional circles grew. Fellow biologists asked for advice. Andy Amman, a fellow graduate student in Iowa, wrote from Michigan about sex ratios in spring populations. Hammy answered within the month. "Pretty good for booming time!" he scrawled. Figures from last fall's hunt, he warned, might be distorted. The hatch had been late, the opening early, most of the birds were molting, and the "moth-eaten critters" that flushed were likely only part of the total number. "What do you make of that? Me, I'm dizzy.

And it looks as though we're going to run into the same late hatch and early season business this year: wet and cold so far, with an inch of snow down here yesterday and the Lord knows what in the north."[5]

Then Fran got notice of her election as a corresponding member of the German Ornithological Society, in recognition of their postwar help to scientists. Hammy was elected honorary member of the equivalent Hungarian group. She quoted the letter from Budapest, complete with spelling, that inquired, "Dear Frederick and Frances Hamerstrom. Are you brother and sister? Twins? Or are you hausband," she smiled, and spelled the word out, "and wife?"

Such satisfying achievements masked growing local tension. Hammy had seen that Vilas Waterman, head of the Bancroft Sportsman's Club, would need special handling from the very beginning. His first complaints came in 1950. "We . . . are beginning to wonder if we amount to anything as far as . . . the bird program is concerned. . . . I understand that you fellows held a meeting . . . with Woerpel in regard to the bird program for Portage County. Where do we fit? Which Club pushed the planting of feed? What bunch of hunters in Wisconsin is more interested in the welfare of the old prarie [sic] chicken than our Club?. . . . Our farmers and . . . Club should be consulted and at least be made to believe that we are a small part of the program."[6]

Waterman, chairman of Buena Vista township, milked thirty-four cows on a small acreage. When the cows developed Bangs disease and had to be sold for slaughter, he became a field man for one of the out-of-state seed companies. Kindly, helpful when sickness or trouble came, he was admired by neighbors. He was outspoken; he could galvanize a crowd. His son Connie, who became an employee of the DNR, recalls going with his father to conservation committee meetings. Connie believed that the Hamerstroms, completely devoted to their work, never understood the effect of their lifestyle. Many residents simply could not understand why they lived so differently.

Fran, the first woman game biologist hired by the then conservation department was "doing a man's work" in a community with sharply and conservatively defined roles.[7] It was not a matter that men were often stronger and could, supposedly, endure more hardship. The usual path to the field was from a strongly male hunting tradition focused on preserving huntable populations. This woman was doing work that didn't even increase game numbers! Her pronounced Boston accent did not help, nor did her climbing trees and other odd behaviors. Fran saw it another way:

We accepted the farmers' ways, and didn't realize that others were not equally accepting. I remember old Mr. Swantek, his feet on a cassock, reading the Bible while Mrs. Swantek milked and cleaned barn. Everybody was happy. I went barefoot and wore awful clothes; we had that snobby way of talking and all those foreign visitors. When our elms died, we left those huge, gray, dead trunks standing. That raised eyebrows. Once, to catch a marsh hawk, I put vegetation on all the fence posts on a field near us except the one that a hawk usually perched on. Suddenly I saw a parked pickup, watching. I ran to explain, but it roared away—another myth in the making on that mysterious country telegraph. "Know what those chicken people are doing? Putting flowers on top of fence posts. Saw it myself." Everything got exaggerated—Os and Mary became our live-in maid and butler![8]

Pheasant hunting was another flash point. The wisdom of that time was that hunters would be served by releasing pen-raised pheasants in likely hunting areas. Those hunters quoted an early optimistic prediction of department personnel that prairie chickens would be back "in good strength" within four to five years.[9] They had ignored the accompanying message—that improved chicken populations would replace stocked pheasants. Pheasants and chickens compete for nesting range; such competition, in the face of the low in the population cycle, became more serious as agricultural development surged.

Early in 1953 Hammy asked for and got a department order to game-farm managers to stop sending pheasant chicks to "sensitive areas like the Buena Vista." Waterman agreed. In May 1953, the local game manager found that Waterman had countermanded the directive, and the Bancroft Sportsman's Club raised money to buy chicks from private sources to "plant pheasants where and when he [Waterman] pleased."[10] Then he ordered his bluegrass crews to give no brood information to either of the Hamerstroms. It was a declaration of war.

Next came the fox bounty fiasco. The Hamerstroms, unaware that part of a town chairman's income came from trapping foxes for the bounty that was paid at that time, earnestly tried to convince him that predators are a necessary part of the ecosystem. In response, that chairman circulated a petition to the governor, asking that the Hamerstroms, like the foxes, be removed. Waterman and a few others spent much of the fall of 1953 circulating it.

Fran knew what to do. "I got the petition. Some of the people who signed it were our friends. They hadn't read it. Or they still held that fox bounties would be helpful."

"How did you get it?"

"I just drove down to the governor's office and asked for it."

The petition had no effect; its numbers and scattered signatures—only twelve of which were from Plainfield—carried little weight.[11] Indeed the respected county representative to the Wisconsin Conservation Congress wrote a letter decrying the attack by a "small vociferous group of malcontents."[12]

Perhaps the Hamerstroms, "gentlemen hunters" themselves, misjudged area sportsmen. For example, when Hammy recommended that the Buena Vista, the site of the best chicken hunting in the state, hence subject to the most pressure from hunters, should be closed to prairie chicken hunting, many of those sportsmen turned against the project. No polite, reasonable explanations helped.

Their beliefs were well ahead of their time. Habitat and ecology were new concepts, social controls, in a time of rapid change, seemed threatened by their iconoclastic ways. A colleague pointed out that, "He walked into an area where poverty and drainage district disagreements had already created a lot of negatives. The Hamerstroms were the lightening [sic] rods."[13]

It was a far from ideal climate; troubling to Hammy who, like Leopold, had believed that the individual farmer was the cornerstone of conservation design. That belief, coupled with his analysis of their data, formed a concept of a new kind of reserve, one he had thought about for a long time. Grange had warned, presciently, in 1947, that "the lay-out for a successful prairie grouse management tract involves the systematic checker-boarding of a large acreage with patches of [many and diverse] . . . farm crops, grass, marsh, berry uplands, burns, trees, shrubs and groves. The exact arrangement . . . is only the starting point. Creating the desired . . . pattern is management, and both hard and costly."[14]

As Hammy walked the marsh, he imagined Grange's checkerboard fitted to the human and bird populations and the Buena Vista landscape. Existing bluegrass acres and small working farms would serve as the core, which he would supplement with scattered parcels of managed forty- or eighty-acre leased or publicly held land. Exploiting the vital edges of farmland would leave considerable land for human use. More units of management could be added as needed and available. It would be reasonable, economical, and practical.

At that time, with only $5,000 in department funds available, leases seemed preferable to purchase. Even then, however, the department land appraiser, Eugene Parfitt, proposed buying sixteen thousand acres at $25 per acre. "Land is cheap, good leases hard to negotiate," he said.[15] "Parfitt was right!" declared Fran decisively as she read this phrase. He saw danger in seed company troubles, in the movement of Michigan muck farmers into the area, and especially in a beginning operation to fatten beef cattle—exactly the kind of profit-rich possibility that would cause farmers to terminate leases. Fran and Hammy became convinced that they must locate, buy, and develop scattered acreage.

On a cold December day in 1953 when Hammy heard from Dick Deerwester of Ansul Chemical Company in Marinette that the Wisconsin Conservation League might be able to buy a property that Carl Hakes was willing to sell, his answer was swift and heartfelt. "It might be the turning point for the whole Grouse Foundation idea. Your letter is the most encouraging thing that has happened in a long, long time." He was careful to suggest a "not over-generous" price, which would inflate already rising costs.[16]

The market for grass seed shrank as Danish and Dutch seedsmen cut prices below the cost of production on the Buena Vista area. Seed companies began to cut their acreage on the marsh. One by one over several years, Mangelsdorf, Peppard, Rudy-Patrick, and Sumner seed companies phased out their operations in central Wisconsin. Meanwhile, with land values boosted by irrigation's promise, Hammy's assessment of the size of the necessary bluegrass core enlarged substantially from his early projection of a six-hundred-acre area.

In February 1954, the Wisconsin Conservation League's purchase went through—coincidentally, unfortunately, with that shrinkage of grass seed acreage. A second private purchase escalated public fears. Conservationists Dorothy and Gordy Kummer stopped by the

Hamerstrom home, simply to say hello. Hammy spoke of his current difficulty: a key sixty-acre farm was about to be sold—and the booming there would end.

"What will it bring?" asked Kummer.

"Not much, surprisingly," said Hammy. "Land has gone for about $17 an acre here, so these small parcels are attractive to hopefuls who think they can 'go farming' without a large investment." Dory Kummer, granddaughter of the founder of Schlitz Brewery, moved in Audubon Society circles.[17] She glanced at her husband. "Sixty-three acres—that won't even amount to $1,500!" she observed. "Maybe we could buy it and make it a reserve."

And they did. What had seemed a modest idea of private purchase of some small parcels to be managed by the conservation department suddenly became a specter for the public: a state-owned marsh. The researchers, already objects of suspicion, were introducing the new idea that publicly managed land be distributed throughout an agricultural population.

An illustrative 1954 map shows how seventy-seven such scattered sections can infiltrate a sample forty-six thousand-acre management area. In contrast, a block would limit the zone of prairie chicken influence to five sections and their immediate surroundings. Fran's later statement of the principle was clear: "Leaving the land be is for specialized stable environments. They are relatively rare. As a rule we need to manipulate the land."[18] Hammy continued, " . . . trying to convince our Department that for chicken and sharp-tail management, land purchase or lease should be . . . [of] a number of relatively small tracts scattered through an unmanaged matrix."[19] Then, in August, some land sold for $40 an acre. "Too steep," said Hammy.[20] At that price, they could not afford the amount of acreage they needed. Still, he intended to move ahead.

The department hesitated, disappointing Hammy, who saw the boom resulting from changing land use escalating land prices in tandem with habitat loss.[21] Eventually Hammy brought not only his agency but also the all-important Wisconsin Conservation Commission around. Unexpectedly, he found an opportunity: an acquaintance, a local farmer named Pratt sent him a handwritten note on a small, torn piece of paper.

Dear Mr Hamerstrom

Not having received any communication from parties you mentioned I am intruding on your time and more. the Probate proceedings are concluded next week and as I am in need of some money for insured expense I have decided to name you a price of only $22.50 per acre. And with the saveing in fence construction it would bring the price to at most $20.00 per acer. Pleas advise.

Hammy turned to enthusiastic supporters beyond the local communities. He wrote Deerwester directly; the League had no available funds. He went to see Pratt in late September, then suggested confidentially to Deerwester that he and Fran could lend the League $600 to ensure getting the land. They had "to keep our names out of it . . . because it would be bad strategy to have it supposed locally that we are rolling in dough and can buy land ourselves. I'm sorry we can't give it outright, but at least you can use it as needed and repay it when convenient, without interest. . . . The Pratt land is too good to lose."[22] It worked; they got the land, established the precedent of purchase, and encouraged the concept of a foundation. [23]

Then the Wisconsin Society for Ornithology, mobilized by the sharptail battles, came through. Over a thousand letters went to members and friends in 1954 urging contributions to a Prairie Chicken Survival Fund. A news release to thirty-nine state papers followed. Joe Hickey increased the momentum with a letter publicizing the Kummer gift of a fine sixty-three-acre block of grassland. He emphasized the moral correctness of the action, and that the "best scientific brains available on the subject" had created the plan.[24] Support continued to grow. When Madison high school principal Paul Olson, a far-seeing, well-connected man in the Dane County Conservation League became involved, the pace quickened. Olson reminisced:

Enter the Prairie Chicken Foundation ( . . . the legal entity for executing leases, serving as a tax umbrella, receiving gifts or grants holding title in fee simple to donated land.) . . . An effort had to be made on the Buena Vista or the chicken would be only a memory. That moment was the spring of 1958. Gene Roark, and I think Buzz Holland, took me up to the Hammerstrom's [sic] to see the "Booming." . . . Here was something so grand, so exciting, I could put my restless energy to work. Furthermore, I had children . . . and oh, God, how a man wants to leave a heritage.

I went to a Dane County Conservation League board meeting . . . and asked the boys for $200 to start a "movement" to save the chicken. We could buy a piece of land (forty acres) for $800, so $200 made a down payment and I launched a newsletter . . . to stir up interest. The next decade was the time of great growth. Both [Senators] Proxmire and Wiley helped us get IRS tax exemption. A radio show or two (Wilbur Stites's *Wisconsin Out-of-Doors*) and the newsletter and we accumulated enough in 1958 to pay off the $800—we felt we had done well. Little did we dream that by 1976 we would have over five thousand acres and, at today's market, well over one million dollars of market value. . . . I also impressed Bill Sieker, a lawyer with remarkable devotion to "decent" causes, who works for us for nothing . . . [and] examined abstracts and drew deeds for the Prairie Chicken Foundation . . . on mostly little pieces—a forty, an eighty, sometimes bigger.[25]

Paul Olson recalled one of the results of the Hamerstrom television exposure. "William Pugh of Pugh Oil Company of Racine called me. He said "I'm going up to Whiskey Rapids (Pugh's name for Wisconsin Rapids, twenty-five miles northwest of Hamerstroms) this afternoon. I'll meet you at the airport." We got maps and went on up and spent the night. That was the beginning of something big."[26]

It was indeed. Pugh loved the experience. Dressed as a farmer, he often visited them. "His overalls were always clean and pressed," Fran reported, "and he kept coming, in spite of being a severe diabetic and having episodes. We kept fruit juice in the house for him." He tried to give Fran $5 for the night's lodging. "We don't take payment from our guests. But, if you'd like to give money for the prairie chickens, it will be put to good use." Over the ensuing years Bill Pugh gave over $100,000 to the Foundation. "If I wanted money in the bank," he said, "I'd put it in the bank. Prairie chickens don't boom in the bank." Later, he gave between $7,000 and $12,000 for prairie chicken research to the University of Wisconsin–Stevens Point each year until his death more than ten years later.[27]

Another stroke of good fortune came in 1961. Dory Vallier went on to meet Bill Sullivan, another member of the National Audubon Society in New York. "It began in the King Cole bar of the St. Regis hotel. . . . " Sullivan recorded. Dory told of the plight of the Wisconsin prairie chicken and declared that they must help.

"Help the prairie chicken?" responded Sullivan, in a now often-quoted exchange. "Hell, I didn't know he was in trouble!"[28] But the

Society of Tympanuchus Cupido Pinnatus that Sullivan founded was to play a major role in the success of the Hamerstroms' plan. When Sullivan learned that the Wisconsin Conservation League, down to only a few members, had voted to disband he called Paul Olson. Fortunately, Olson had not yet filed the required resolution: its tax-exempt status was extant. A simple vote to rescind the resolution to disband, a new name, and the election of new officers, gave birth to the society. Sullivan was its first president. By 1966 the membership in the new Society of Tympanuchus Cupido Pinnatus included 175 corporate officers, 82 business executives, 45 lawyers, 27 doctors and Governor Warren Knowles (who served on the board for many years).[29] Indeed, Fran loved to tell of the time that a reporter said to her "Fran, all I can see in this room is money. If something should happen to this building this moment, every business in Milwaukee would come to a halt."

The Hamerstroms described their approach to the Society's organization in their high-spirited 1961 report. Their structure provided "a whiff of laughing gas to ease the pains of cash extraction." Bylaws, with gentle irony, restated Robert's rules of order. Number five, for example, "Ruffled Feathers" set up procedures for withdrawals from membership; Number six ("Day of Reckoning") described a hierarchy of giving. Old Pros were the scientists and area managers; "Big Boomers," members under forty who contributed at least $5 per year; "Akermakers" donated up to $25; "Big Operators" made forty-acre purchases possible—and, if they liked, got their name on a commemorative sign. A council of chiefs (Tympanuchus, Pinnatus, Cupido, and The Quill) were the officers, in normal order.

"Good works don't have to be done in a sepulchral atmosphere," said Hammy, in one of his memorable one-liners.[30] Gala cocktail parties followed blessedly brief business meetings—all regularly reported in the society columns of Milwaukee newspapers. A newsletter (*BOOM!*) impressed readers "with the clever wording—for some reason every page seemed to ring of Fran."[31]

For some, membership meant real involvement. Well-connected lawyers and business people in Milwaukee or Madison brought maps and documents involved with land purchase, news of available parcels, and notice of available bulldozers or mowers up to Plainfield. Most members, however, simply visited the Plainfield headquarters from

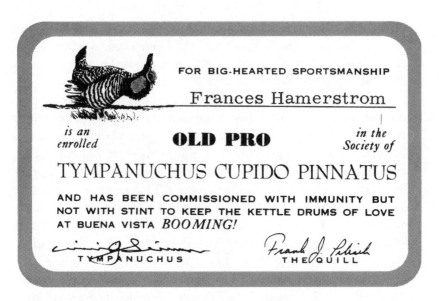

FOR BIG-HEARTED SPORTSMANSHIP

Frances Hamerstrom

*is an enrolled* **OLD PRO** *in the Society of*

TYMPANUCHUS CUPIDO PINNATUS

AND HAS BEEN COMMISSIONED WITH IMMUNITY BUT NOT WITH STINT TO KEEP THE KETTLE DRUMS OF LOVE AT BUENA VISTA *BOOMING!*

TYMPANUCHUS          THE QUILL

Fran's Society of Tympanuchus Cupido Pinnatus membership card

time to time and enjoyed the lively social occasions (euphemistically called "business meetings") held first at the University Club and then at the Uehlein Performing Arts Center. Years later, an involved member recalled the atmosphere and stratagems: by-laws required that formal actions, mostly about money, must appear in the minutes. Since the parties were neither formal nor businesslike, the then Quill Jack Pelisek made sure that he and Bill Sullivan Jr. (Bill Sr.'s son) would make and second certain motions in a way that would guarantee unanimous passage. Nothing should interfere with the camaraderie.

There was one formality: members addressed *both* Fran and Hammy as Dr. Hamerstrom. When asked who responded when a query was so directed Fran replied, "It depends on the tone of voice."[32] That approach, the cause, and the Hamerstrom alchemy built a significant membership.

It was just Fran's cup of tea. She loved dressing up and mixing with people who reminded her of her Boston youth. Hammy, distinguished and courtly, won respect and admiration. As one visitor said, "Fran tries to charm people; Hammy does it." A snowball effect ensued. Society members responded to the invitation to observe the booming.

Some brought friends; international visitors came, and the cadre of enthusiasts grew. Dedicated conservationists like Olson were energized: "He was just such a decent chap! You simply had to help him all you could." That help made 10,770 acres on the Buena Vista marsh safe for chickens by 1969. A jubilant Hammy thought it spectacular, but he credited others. "There is magic in the names of the distinguished "old timers" who were among the first to take up the fight for chicken survival: for example Dick Deerwester, Hugo Schneiders and Clarence Searles who turned the last funds and energies of the old Wisconsin Conservation League to saving chickens and then disbanded; Paul Olson, Bill Pugh and Bill Seiker who formed and still guide the Prairie Chicken Foundation of the Dane County Conservation League; and the shapers and founders of our own Society, Bill Sullivan—Tympanuchus Himself—Dory Kummer (Vallier), John Best, Nondy Hinrichs, Jack Pelisek."[33]

"We were simply the catalysts," he would say. But once he casually remarked, "Damn little land would have been bought on the Buena Vista marsh without our fine Italian hand." Fran, knowing that the two of them created as good copy as the chickens, made room for every reporter, photographer, and writer who asked to come. Critics of their lifestyle had failed to see that it would add to their appeal.

They drew all kinds of people to them: bird lovers, the curious, the scientific, and the dedicated conservationists. I was none of these, but we were neighbors and we liked the Hamerstroms. By the mid-1960s, I had become aware of the importance of the count at booming and understood the reason to stay at the house: a single latecomer could prevent timely deliveries to several scattered blinds. I asked Fran if she could fit me in again. "I'd like to see some action," I said. She gave me a speculative glance. "I'll see what I can do." When someone canceled, she called. "I'll put you in with one of our regular observers," she said. This time I savored the experience. The participants—a biology student and his bride from Ripon College, a teacher from Wautoma with some youthful helpers, and a couple from Milwaukee—were all enthusiastic, informed, and even exhilarated.

Hammy's briefing at 7:45 P.M. sharp was authoritative and explicit. I took notes. He said, "This is a research project; please help us record what happens on each of our booming grounds. Their size may vary;

some are smaller than this room, some much larger. Each cock's territory will be perhaps twenty feet square; we hope you will notice exactly what part of the ground it occupies." He described the staring and sparring matches in which the cocks engaged: they might flutter straight up in the air, sometimes as high as five feet. "They will fight for their territory. Since they all want to be where the hens are, you'll see more activity when the hens turn up. Just like people," he added with the hint of a smile.

"You'll need to be a detective to figure out the number of hens. They run the show. One day a hen may seem casual—sitting, scratching lice, even joining with other hens for a sister act. The next day, she'll come in and act a little more noticing, more and more aware. She'll make short runs, shy away from the cocks, but not very far. By the time she's been there several days, she's ready. Then she makes her choice.

"We try to keep the grass mowed on the booming grounds. This helps the birds, who want to be seen, and makes your job easier." He held up a small object. "Our bands can easily be seen with binoculars from one hundred yards. Record the number, color, and position on which leg of each band." He displayed a spotting scope but advised a good pair of 7 x 35 binoculars for first timers. "Only one scope to a blind, please."

"The same prairie chickens return to the same booming ground year after year. Two or three cocks do about 80 percent of the mating though the whole crew is there each day. The hens seem to know these important cocks. Genetically this has interesting aspects. Watch carefully, make a map of positions and activity, get band numbers, and keep track of times of arrival.

"In the morning we come down, boots on and handkerchiefs in pocket. Nobody," he said with awesome firmness, "goes back upstairs. We eat here, then go to that table over there and pick up 'scopes and clipboards—one to a blind. Be settled in the blind before the birds arrive, or you'll flush them." He paused. "If you lose your way to the blind from the road, get down on your knees and sight against the horizon. No flashlights, please. Look for an unnatural object, square rather than round. That will be your blind.

"Stay as long as there are any hens on. Sometime after the hens have left, the cocks begin to simmer down. When that time comes—about 7:00 or 8:00—just dump the blind and flush the birds." He drew a breath. "Now for identification." He held a remarkably lifelike stuffed specimen—a "skin"—in front of him and continued.

"The cock is bigger than the hen and has longer flange feathers on his neck. There's no mistaking him when the neck tufts and tail go up and," he pointed to the orange colored sacks on the neck, "the bellows inflate." He picked up another skin. "You may see a sharp-tailed grouse, a bird like this that is plentiful in the barrens up north. Observe the shape of its tail." His fingers touched the feathers, and then drew an exaggerated point in the air. "They dance in the spring too, and the species interbreed, producing hybrids. Distinguishing these is harder, but important; we'd like an accurate count." He turned the skin slowly around. "Any questions?" He waited. "Come up and handle the skins if you like."

He made the procedure clear. "We go bed early to be ready for the morning; Fran will show you where you'll sleep." She gave me Elva's room. After a walk, flashlight in hand, to the drafty outhouse, I was glad to crawl in to a musty bed, heavy with blankets. Almost immediately, it seemed in the black dark, a resonant voice called. "Breakfast!" Stumbling into outer clothing, I smelled coffee, found bacon, eggs, and oven-broiled toast handed out by a cheerful Fran.

"Get your boots on while you're eating," urged Hammy. "We've got forty-five minutes to get into the blinds." My businesslike companion was skillful, releasing the anchor ropes and tipping up the four-by-six-foot structure, readying scope, small clipboard, and positions in a practiced set of maneuvers. He did the recording; I absorbed the light as it took the darkness, the sounds, the crisp, clear, empty air. I counted as the booming call commenced: one-chimpanzee, two-chimpanzees, three-chimpanzees, four . . . until the reverberations died. My companion was busy, alternating observing through his scope with jotting band numbers. I watched some dozen birds, including a cock who stared at another and stamped his feet so rapidly that they became a blur as the other cock approached, and then, with tail spread, jumped and whirled half around. I saw a brief encounter where the two cocks actually met, in air, in a flurry of striking feet and a few floating feathers—and then started calmly pecking the ground. I saw a hen wait, calm but expectant, for a cock to mount as she stretched her wings wide. "Sometimes," whispered my companion, "the cock will actually bow." After forty-five minutes or so, the action slowed down, and gradually the birds dispersed. Still clumsy with the binoculars, I asked my companion, "How did you ever get so good at reading those bands? I can hardly locate them." He eyed me dispassionately. "It's harder with glasses," he replied, "and it takes practice."

When we returned we found Hammy at a small table, his hair neatly combed and his Pendleton shirt buttoned at the neck. A half-empty cup and a percolator stood beside him; his boots and a high color revealed that he had been outside. Reading glasses low on his nose, he inspected the notes of a dozen observers that would be coded onto permanent record sheets. While we waited he debriefed the man and wife from Ripon in his low, musical voice. "Let's see—you were at the Walt Disney Blind. There were nineteen cocks and three hybrids there last week, but no hens. You say you saw three banded cocks? And two hens? Good!" He jotted marks in columns.

"Did you see any sharp-tails?" He nodded at the negative answer. "Still some hybrids?" He accepted the confident, "Three, I think." This young man had been booming before. "Were there any successful copulations?"

"How," blurted the young wife, "do you know if they are successful?"

No one smiled. "It was probably successful," he explained, "if the cock mounted squarely after the hen invited. Then, if she made little runs, roused and left the booming grounds within five minutes, you can be sure she had lost interest in mating. Good question. Any band numbers?"

Hammy had been up since 3:30 and had stoked three stoves while Fran crisped the bacon she had half-cooked the night before. With all blinds staffed, he took time to cruise the marsh, listening for possible rejuvenated booming grounds and reveling in the sunrise. Our reporting was methodical, thorough, and took no more than five minutes.

When Fran appeared, handsome in outdoor gear, I asked if I could help. "Do pass these sandwiches around." I did so, watching Hammy stow his recording forms in a folder, set the specimens on a shelf, finish his coffee, yawn, and come into the kitchen. "May I help with the

Hammy in the early 1960s, studying an oberver's map of the booming ground activity assigned to him that morning. *Sheboygan Press*

dishes?" I ventured. She refused; I was by now comfortable with asking her if Hammy always did the briefings. "Usually," she replied, "but we often work together on checking the field notes and coordinating them with maps of each booming ground. The two of us can finish in an hour. And the notes help us judge who can best handle the difficult assignments."

Leaving, I heard him ask if there were sandwiches left. Fran admitted that in spite of eating four hearty meals a day, they each lost up to ten pounds during booming season. "We needed two mornings in every day or ten days in every week to get through the necessary work." The grace and courtesy with which this routine had been managed through years of fluctuating populations and personnel was remarkable.

Booming was hardly the only reason that people came to see the Hamerstroms. Kip and I loved going up there. We'd drive past farm fields, then through oak and maple woods, and turn in past the old windmill and classic barn. We'd walk the rough path past their imposing woodpile and step up the uneven stonework stoop with wildflowers and crimson cut-leafed geraniums spilling over its edges. We'd pass the pear-shaped tin bathtub hanging on the back wall and push open the unlocked back door.

Then anything could happen. They accepted us as we were and naturally included us at whatever their level then was. Hammy, in tailored corduroys, a muted brown plaid western-style shirt, and a bolo tie, might turn from the dishpan or the pump and say, "Fran's on the side porch with her heron," and we'd find her admiring its remote, head-cocked stare. Or she might introduce a visiting European ornithologist, a museum taxidermist, or another neighbor. She might lead us outside to visit the broody female eagle, Chrys, or to watch the flight of a hawk and its return to her low whistle and the reward of the mouse carcass she pulled from her pocket.

The pump house entry to the house led past a cluttered storeroom, snowshoes of several sizes hanging from the rafters, through the screened door into an extraordinary chamber. A dingy refrigerator, a vintage green electric stove and a wood-burning "garbage burner" identified its primary purpose, but it was the wash room, too. A tin dipper marked the clean water pail; it stood on the floor handy to enamel basins on a low rustic bench. The dump bucket stood at the other end. The top of a waist-high wooden cupboard served as a work surface, carving board, dishwashing station, storage space, and counter. Bread and pies took shape there; wine bottles and spices stood in mixed order at its back. Utensils hung from a shelf above it, and the nearby open door of a superb unpainted country kitchen safe exposed tall piles of assorted plates and cups.

At one September luncheon Hammy greeted us with unusual gusto. "Ah! the Cornelis! Fran has turned your mushrooms into one of her famous pies. They are ready." Hammy loved pie. We showed him the huge mushroom we had found and wondered if we could eat it. "Ask Fran," he said. She appeared in short shorts, a Mexican blouse, and huaraches. She regarded our mushroom severely. "We have forty-three species of edible mushrooms here; we eat them year round." She got out her pamphlet. "Let's try *Leucoagaricus,*" she announced as she turned the pages. "It is a *Leucoagaricus,* but is it *naucinus? procerus?* Or *rachodes?*"

It was *naucinus*. "Not recommended. But we're safe with your lovely honeycaps! I'm drying some, and here are the rest. Get a plate and silver, there, under the chimney. Pie's cut, I made six this morning." A battered twenty-quart pressure cooker held potatoes. "Help yourself, and take your plate into the table."

Plates in hand, we moved into the central room, dominated by an old Royal Oak 2000 stove almost six feet tall. Fran scooped up a stack of paper from one of the chairs and added it to piles on the sofa. A rank of large manila envelopes, labeled and loaded, waited for attention on the back, and what appeared to be a manuscript had been swept into a casual heap on the floor. We sat in four fine antique Windsor chairs around a classic oval cherry table.

On the back of the fifth chair perched a great horned owl. He swiveled his head to inspect us, the table, and, minutely, his claws, then sat in silence, his great golden eyes unblinking. The table served as an adjunct desk: a Dundee marmalade jar held pencils, pens and a letter opener wound with rubber bands, a small flashlight. Hammy sat at the head; Fran nearest the kitchen door. "Beautiful sterling, Fran," I said as I hefted a massive fork. "Grandmother's" she retorted briefly, and returned to her pie. Kip regarded the owl on the chair back next to him.

"Hello, Ambrose." There was no response.

"Lovely crust, Fran," I murmured. "Lard, I suppose?"

"Bear grease!" she announced triumphantly. "It makes the best crust."

"Where in the world do you get bear grease?"

"A friend brings it to us. Have some more pie." We did; and we talked of the risk in eating mushrooms.

"I'm sure of my species," said confident Fran.

"But we start with small helpings," countered Hammy.

Strong coffee appeared and the pace of eating slackened; talk livened, and then stopped as the loud whining roar of a low-flying plane intruded. "Do your neighbors spray?" asked Kip innocently.

"Don't mention spray planes!" growled Hammy. "I'd like to take my shotgun to them when they come roaring over us at dawn!"

"I hope you won't!" said Kip. "Look what happened to Czlapinski." This man had received a significant fine for firing a gun at a spray plane.

"Czlapinski should have gotten a medal!" replied Hammy indignantly. Kip, an irrigation farmer himself, explained: plane-applied
insecticides are a way of keeping tractors from compacting the soil.
Hammy heard his defense with polite skepticism; Fran reacted vigorously with statistics on the mortality of nesting songbirds. Ambrose
grew restless, began to sharpen his beak on the back of the chair.

"He's bored," said Fran. I, thinking to distract him, offered a
crumpled paper napkin. He grasped it in an eager claw and bent to
rend. The reaction was electric. Hammy, always soft-spoken, drew himself up, fixed Ambrose with a fierce stare.

"Ambrose!! NO!" he thundered. The owl froze, ducked his head,
and curled his claw tightly around the napkin. Fran swiftly proffered
a fountain pen; Ambrose reached for it. Deftly she removed the napkin. Ambrose held the pen in his beak, a bit sullenly, then dipped his
head, transferred it to his claw, and regarded it bleakly.

"Paper is bad for owls?" I inquired, startled and abashed.

"Owls" said Hammy, gently, "are bad for paper!" Then, after taking his plate and cup to the kitchen, he retired through a wide arch
draped in weighted fishnet to his office. Birds—which often flew free
in the house—must not enter, and an unspoken convention kept visitors and friends, too, outside the barrier when it was closed, as it almost always was.

The room spoke of a full life, a unique life, a life with priorities
felt and clearly followed. Every piece of the furniture was functional
and cherished. The graceful Windsor chairs were fragile. "Don't rock
back!" she'd warn guests. Decoration was eclectic. Crude shelves
in the chimney cupboard held a boldly executed wooden plaque of a
prairie chicken, two carved wooden game cocks, duck decoys, and a
pile of dust-covered journals. A dozen Mexican shell necklaces, looped
on one side of the chimney cupboard, might call forth tales of winters
spent among the Seri Indians in Mexico. Jumbled hawk jesses evoked
visions of birds of prey tethered to a perch.

Unexpected beauty greeted the observant. A fine, muted-red kilim
rug hung over a bench piled with mailing labels, Indian baskets, shells,
a prescription, a carved wooden shark, a small wreath, and plastic
bags with mysterious contents. A framed color photograph of Fran
holding a spray of orchids and smiling up at Hammy told of travel. The

walls of sand plaster, a smoke-shaded cream color with fine cracks and smudges, displayed a watercolor of a young heron by Mallik, a pencil rendition of winter-iced oak branches from which a goshawk gazed, and a large watercolor of an owl whose golden eye surveyed the glacial kettle that broke the sweep of the winter expanse of an old prairie cemetery.

"That's a Jonathan Wilde. It's Hammy's favorite," Fran volunteered as her look followed my gaze. Bookshelves filled with well-worn volumes invited browsing. One shelf was tightly packed with colorful paperbacks and hard covers with bright jackets, each with HAMERSTROM on its spine. Two tiny stone owls guarded these volumes. I picked up one of the little owls, asked about it.

"Oh, people have been giving me owls for years. I thought those deserved display."

Stacks of *New Yorkers* lined one wall, their covers annotated in bold black: "*Hammy—page 79,*" or "*Fran—see 82.*" Hammy saw my glance. "We can't keep up with them, the articles are so long. Still, some mustn't be missed."

Hammy would listen to Kip or to me, often responding only with a resonant and encouraging "Hm-mm," or another quiet question. But he loaned me *Freethought Today,* to which he subscribed, and a booklet about the Roman Catholic Church's financial support of the radical right-to-life movement. I wondered if it was overwrought, so gave it to a Catholic friend for review. She judged it, sadly, as "probably accurate."

They had particular firm opinions on which we disagreed. One concerned "grammar." I hold that yesterday's proscriptions become today's practice. They, on the other hand, were traditionalists: he winced when he heard certain locutions. "Like and as!" he would groan. "You can't imagine the number of graduate papers I read where the authors simply don't know the difference! Appalling!" I never heard him correct anyone, but Fran did, forcefully. "Different from!" she'd insist, frowning. Certain opinions were fixed: no argument about the failings of the so-called science of eugenics made the slightest impression. "Surely you recognize that breeding is essential to improving the race!" he said, almost huffily.

Hammy steered most conversations away from himself, although occasionally he might tell us of an early experience, such as the time—before the advent of modern sound recording equipment—that he helped a friend attempt to record the calls of sandhill cranes (at that time wary birds indeed). Hammy helped set up the heavy equipment in a blind, retreated to an appropriate distance, and waited a long time. No crane called. Would an imitated call initiate a response? It did not. Eventually he called out, "Might as well head for home." As he approached the blind, a wildly exited man leaped jubilantly out, waving his arms and shouting, "I got it! I got it!" He had recorded Hammy's call.

Hammy shook his head. "I hardly had the heart to tell him," he confessed.

Conversation never flagged; they were stimulating, empathetic, interesting. He and Kip commiserated with each other at income tax time; we shared distress about Watergate, the use of Agent Orange in Vietnam, and any kind of entrapment. But he never joined in the rare occasions when Fran and I descended, as he clearly thought, to repeating local news. "Woman talk!" he would exclaim as he disappeared through the fishnet.

Such a reception—different in detail but not in spirit for each visitor—made visits to Plainfield a tradition for some, an obligation for others. Madison photographer George Socha and his family spent a couple of April weeks there in 1957, 1958, and 1959. One year Socha helped with the booming for a full six weeks, often driving back to Racine after each early morning's observations and returning to Plainfield late at night after taking care of his own business.

Children were welcomed. Businessman Harry Croy brought his family and then his Scout troop. They arrived as Fran, in hot pursuit of a red-tailed hawk, had cut through the sod of a just-thawed meadow. By the time a farmer pulled her out, her wheels had damaged the field "beyond belief." She promised to have it fixed by the next noon and, thanks to the fortunate arrival of Harry Croy's Boy Scouts, it was. (Fran, in her way, augmented the story a little in *Strictly for the Chickens*: she shrank the Scouts into Cubs, "gnome-like little children" with little hands that patted the sod back in place.)[34] Croy also brought his father-in-law, who was on the board of Carroll College. He was so impressed by Hamerstrom dedication that he arranged for Fran to be given an honorary doctorate in 1961.[35]

Fran and Hammy circa 1960. They made both avian and human guests feel at home. Here, Hammy displays a harrier, and Fran holds a snowy owl.

People visited the Hamerstroms repeatedly, not for the novelty but for the respect they received and the vision of what directed freedom could bring. Paul Olson's son-in-law once spotted a certain banded bird in a certain location. Hammy said that couldn't be—but soon he wrote to say that the novice had been right—the bird had moved. "Characteristic," said Olson. "They were inspirers, they changed lives. We went up every year in the '60s; my daughter never forgot the goshawk she held in her hand."

John Emlen, professor of ornithology at Madison, brought his family and his classes. His twelve-year-old son, Steven, came back from his blind saying he had seen a brown pelican. "Can't be," declared the observers. "Too far north."

Hammy quietly inquired, "Where was it?"

"Right at the edge of the booming ground."

"How would you describe the color?'

"It was brown, not a dark brown, and big. Almost as big as a goose." Within a week, Fran saw it at the same blind. She wrote

Steven, now a well-known ornithologist, that very day, as his parents warmly reminded her in my presence, some thirty years later.

Dr. Emlen spread the word about a "superb introduction to prairie chicken biology and conservation." He valued the demonstration of the importance of note taking, the precise arrangements for pre-dawn transportation to a blind, and the encouraging tone. They were, he wrote, "sent to bed with a warm pat on the back. After from four to five hours of unforgettable watching and listening . . . all were returned to HQ for a round up of reporting. . . . then [it was] back to the relative boredom of the university classroom."[36] Other professors and high school biology teachers saw the booming as a remarkable teaching opportunity. Auto caravans and busloads of students arrived each spring.

No longer did Fran have to scrabble to find boomers. They devised methods to manage the growing amount of data and occasionally delegated the briefing and the follow-up to a helper. Now they needed to deal with the complications that grew from their innovations.

# 12

# The Prairie Chicken War

*The sooner ecological patterning is whipped into fully usable form, the sooner will wildlife management be ready to cope with the problems of the future. These problems, indeed, are already upon us.*
—Frederick Hamerstrom Jr.

In 1954, when conflicts about conservation were less widely publicized than today, saving the Wisconsin prairie chickens was seen locally as a David-and-Goliath battle: the state against the little person. The Goliaths, powerful DNR bureaucrats from Madison, wealthy outsiders, and those peculiar Hamerstroms, were the villains. Against them stood the champions of the little guys: seed company employee Vilas Waterman, small town druggist Leo Gwidt, and marsh farmers like Harry Isherwood and Tony Palek. And so what the beleaguered conservationists soon dubbed "the prairie chicken war" began.

The "scatter pattern" reserve seemed so logical, so reasonable, and so clearly a win-win situation that the furor took Hammy, buoyed by their early land purchases, by surprise. To a reader today, the idea behind the Hamerstroms' *Guide to Prairie Chicken Management* seems sound. The basic principle was to put underutilized land to good use by spreading a modest acreage of managed land through a much larger area. This would provide over seven times more of the needed edges than would a solid block of managed land. Existing farms would provide cover, some food, and perhaps even sites for booming grounds on suitably grazed meadows; managed acres would address the vital needs of nest-brood cover and winter food.

When the Portage County Conservation Committee asked Hammy to explain the prairie chicken work he cheerily gathered the necessary

181

maps and slides, stowed the projector, and drove off eagerly to the North Grant town hall. He returned, dispirited, late in the summer night. Fran saw his expression, his sagging shoulders. "I thought they were going to tar and feather me and run me out on a rail. Let's talk in the morning."

Breakfast was somber. "I expected a normal meeting to consider closing parts of North Grant and New Hope townships to hunting. But cars were parked all up and down the road, and all around the town hall. It seemed unusual." He closed his eyes, leaned back.

She filled his coffee cup. "Warden Jelich wasn't there," he said. "Gwidt was; Vilas Waterman came in late. Gwidt did most of the talking. He was loud, bitter, and profanely critical of the department." He passed his hand distractedly across his forehead. "I can't give you all of his arguments. The closed area was posted too late last year, though he excepted our project from blame. What had we done or learned that Grange didn't do before us? Why haven't we produced more chickens? Indeed, why didn't we buy up the whole Buena Vista marsh? And why didn't we take better care of the chickens last winter instead of waiting until snow was on the ground to start feeding? 'Everyone' in Portage County saw the inadequacy of our program. I suppose he was talking about that one field we shocked after snowfall."

Fran waited.

"Well, after a discussion of the closed areas, all quite smooth and pleasant, I turned to him. I told him that I had just realized that I had never made clear that the grouse project was not intended for development work. We were to find out what needed to be done by some other outfit. I spoke as simply and as pleasantly as I knew how—I can't believe that I came across as having any chip-on-the-shoulder attitude."

"I'm sure you didn't, darling."

"Suddenly he launched into a furious, completely unreasonable tirade. I can't report it in any orderly way. He was not going to take any more of our nonsense (definitely not his term) about long-term studies. The department passes the buck; Cy makes promises which he then 'forgets.'" Hammy fell silent for a moment. "Waterman was quiet on all the other points, but he joined in on that one. Then Gwidt stormed on: Cy has told Gwidt that if he ever catches me out of the office doing fieldwork, he will fire me; but Ted Hall saw me on the marsh before five one morning, so he supposes I do some fieldwork, after all. Gwidt isn't afraid of me," he smiled wryly, "or the higher-ups, or anybody else, and we can count on him to go on being critical when he feels he should." He paused, sighed. "Anytime."

"It sounds awful."

"It was, until, unexpectedly, he became pleasant again. As we were leaving, he came up to me and announced that he was the kind of person who liked to bring things out in the open. I asked him if he 'felt better' about the grouse project after our discussion. By George, he said he did. Maybe," he paused, "maybe some good came out of it after all. I hope so. It certainly took a lot out of me."[1]

Gwidt's tirade was almost certainly based on misinformation and limited understanding. Cy Kabat might have told a complaining citizen that administrative employees should not be doing fieldwork; critics, however, exploited anything they took as a violation of job requirements or project necessities. Those who feared state ownership of land on the marsh considered rising land prices, the advent of beef operations, and the movement of land-hungry outsiders onto the marsh to be evidences of progress.

The department sent a presumably impartial staff member, Russ Neugebauer, to assess the situation. He was of the "Information and Education" division but turned in a hasty report—based on interviews of "a number" of unidentified local residents. He advised against leases. "We would [have to] set forth specific practices which the farmer would have to undertake." His prescription was education and cooperation, "because of the fact that I believe we can encourage the people living within the area to do those things on the land which would be beneficial to the chicken and at the same time beneficial to the farmers economic welfare."[2]

Neugebauer's facts were wrong, his judgments unsupported, his estimate of costs—probably from farmers—well above Hammy's "worst case" projections. He claimed that only the rare, well-financed large landowner could "maintain a large population of cattle which may lead to deterioration of the bluegrass crop" and maintained that others would delay pasturing until the brooding season was over. He inferred that state help could preserve the bluegrass industry and that conservation committees and county agents would be cooperative. [3] The margins of Hammy's copy bear explosively large question marks and *Nos* at the worst of these superficial judgments. Finally, he recommended that neither the Hamerstroms nor Os should handle *any* "public contacts." Animosity toward them would prevent any cooperation in the area.

Neugeberger's follow-up letter to the Hamerstroms ten days later was not helpful. "I would have preferred to talk the situation over with you folks before the information was submitted." No attack upon "your integrity or your ability" was intended; "within the near future I shall be in a position to stop in and discuss this matter more completely with you. . . . All for the good of the prairie chicken."[4]

Fran, confident of the goodwill of the marsh farmers, dismissed it. "That's all gossip," she said: "The farmers up there treat me as cordially as they always have." They were used to misstatement and exaggeration. An unknowing person, seeing traps in blizzard weather and unaware that they were unset, could assume frozen birds therein. Once spoken, that assumption would spread and the story that traps had been in the field, with chickens freezing to death for two weeks would make the rounds. They had heard about the bands. Three on each chicken leg, it was said, weighed the birds down so they could not fly. Hammy's analogy, that a chicken weighs about two pounds, a band two grams—less than a Hershey bar in a man's coat pocket—reached few.

Underneath all the talk was the fear that state-owned land would take income from the meager tax rolls. Dirt-poor country people thought "all that money" should benefit locals. A native, they maintained, could do the work Mrs. Hamerstrom was supposed to do—if, indeed she worked. She had been seen sitting in her car, on the roadside, *knitting!* Years later, a neighbor explained; passersby saw her as idling when, while timing the afternoon booming she was knitting the German way, fingers automatically feeding the wool to the needles with her eyes on the fields.[5]

Then came the goose gossip. A local resident asked the Heart of Wisconsin Conservation League to consider an available eight hundred-acre marsh farm as a possible goose reserve. Hammy wrote the Horicon Waterfowl Biologist, "I'm sure that the people who live on the marsh would make a terrific uproar if they thought the Department were seriously considering such a thing. . . . And so would I."[6] The idea got no official backing, but the suggestion generated talk. That talk brought another angry meeting, and Hammy put on record the bias and mistrust demonstrated there:

The general tone . . . was one of furious anger. . . . Although there were a few men present whom we know to be sympathetic to the Project, no one spoke in our favor. A number of matters were brought up . . .

1. . . . purchase would mean loss of local taxes. . . . I explained that the Commission would not authorize purchase . . . if town boards objected. . . . [They] refused to believe it.
2. The State will flood the entire marsh. I . . . showed the map of our . . . proposal . . . the ringleader insisted that that . . . was simply an attempt to gain a toe-hold . . . people higher up in the Dept are trying to get the whole marsh.
3. The State will take what it wants through condemnation . . . we have proceeded in secret. . . . I was so rattled by the complete injustice of the statement that I failed to point out that I had described the plan to the Bancroft club two years ago, and that Waterman had given Town Board approval in writing at about the same time. . . .
4. The two parcels . . . privately purchased for chicken management have done "hundreds of thousands of dollars worth of damage" to land values. . . . I pointed out that some thousands of acres have changed hands . . . at about the same price that was paid for the 143 acres bought for chickens. Nevertheless . . . those 143 acres have torn down the value of every farm on the marsh.
5. . . . [Some] were polite. . . . Their cause is to win appreciation for the marsh as agricultural land and they are firmly convinced that there just isn't any room for wildlife.
6. . . . Waterman repeated his theme. "You can't tell us farmers what we can do or can't do on our own land—this is our dunghill and we'll fight for it till we die." He was applauded.
7. There was considerable angry talk of hunter damage. I pointed out that . . . chickens were hunted only once in the last 10 years.

. . . I wish I could report that . . . I succeeded in winning them over. I didn't. They were roaring mad when we got there and roaring mad when the meeting broke up. . . . We have learned . . . [that] the whole thing started as a genuine

fear that the whole marsh was in imminent danger of being flooded. . . . I still
don't know . . . how chickens and the grouse project were drawn into it. I
think it fair to suspect that it was done deliberately, however. Harry Isherwood
. . . was the organizer and leader of [a second meeting] . . . on the 18th; we
were not invited. . . . I am told . . . [it] was equally heated.

In talking with landowners since the 13th . . . we have not found one . . .
who did not talk with us in a friendly manner. . . . At least three farmers have
offered to sell us their entire farms. . . . I'm sure that . . . objections can be
overcome, when the people on the marsh have cooled down enough to be able
to think for themselves.[7]

Hammy continued to be outwardly calm and courteous. When
Mrs. Rickmeyer, a marsh farm wife who believed the stories, sicced a
big brute of a dog on him, "I barely managed to jump back into the
car." A few days later he came upon her on the road, stuck deep in
the spring mud. "Of course I pulled her out." And he publicly credited
the bluegrass strippers for "the enormous help they've given us over the
years."[8]

Ten farmers went to the State Conservation Commission in December where Waterman, speaking, he said, for 156 farmers, called
the area Hamerstrom's Kingdom. "For the amount of money spent on
prairie chickens they should be wearing gold suits." And, glaring at
Os, he complained, "Conservation Department employees should be
working instead of dealing in real estate without a license." Commission members professed ignorance of complaints; he replied, darkly,
that they were not informed on a lot of things. The commissioners,
declaring that no socialist plan would be adopted, took the matter under consideration. [9]

Meanwhile, the demands of the project continued. Kabat, under
pressure to economize, requested a list of management and research
projects and "harvest" results for each year since 1939 with an estimate of dollar savings.[10]

"Listen to this, Fran: 'I see no reason why you cannot prepare these
lists in about one day,'" said Hammy wearily, staring at the letter in his
hand. "It's April. I can't neglect the booming."

"Let me answer Cy," she offered, and then hesitated. "Not today,
students from Madison are coming. They'll need extra attention."

"No, Fran. I'll do it." He moved off into the study, muttering.

"Dollar savings! Fran!" he called from the archway, "what do you pay for those twenty-pound packages of cheese you buy each week for the second breakfasts?"

Withal, Hammy remained optimistic. He maintained that their plan was reasonable; and—in spite of all evidence to the contrary—that people were reasonable. Most marsh farmers were friendly and cooperative. Local residents knew what was intended and responded positively. He asked for positive support from public relations personnel. "Chickens need help, and to get it they need good publicity. . . . too many people . . . believe either that the job can't be done at all, or that we have so few birds left that there is no point in trying to hold them. Both ideas are wrong. The prairie chicken can be saved in Wisconsin if a really adequate management program is started soon enough—which means right away. And the population density on our one best area (the Buena Vista-Leola marshes) is one of the best in the world. Let's not give anyone grounds for believing that the Department has thrown in the towel!"[11]

He spoke to any interested gathering, from Izaac Walton Leagues to the Girl Scouts. The largest meeting, sponsored by the Conservation Fraternity at Central State College in Stevens Point, drew a crowd of 150 people. There he reiterated that complaints arose from "misunderstandings," serious because they threatened the marsh, the only location in the state "where we can be sure that the prairie chicken can be saved." Almost five thousand acres of grass had disappeared with changing land-use patterns in the past two years, while a mere 988 acres had become suitable habitat. His plan would deal with the trend. "Our plan calls only for a scattered pattern of forties . . . limited acreage . . . placed under controlled management."[12]

Newspapers all over the state picked up the story. Letters to the editor featured prairie chickens. Les Woerpel's Christmas newsletter to the Wisconsin Federation of Conservation Clubs spread the alarm and called for a public hearing, with sworn witnesses and an examiner.[13] The Portage County Sportsman's Club passed a supporting resolution. It listed and then refuted "false and preposterous charges" that

the state plans to buy a large portion of the marsh thereby depriving the town of taxes; that private individuals plan to buy large areas and give them to the state; that the state will take land it wants through condemnation . . . that the state intends to propagate prairie chickens on the marsh so that the birds will multiply to such an extent they will fly out to other peoples' lands and draw in thousands of hunters who will ruin farmlands . . . and drive the farmers off the land.

No authorization has been given the department to purchase or accept as a gift any land on the marsh. . . . it would be impossible to flood the marsh under the pattern of ownership or lease set up by the grouse research program which contemplates about one forty in a section . . . the department has never condemned any lands in the state for game management purposes.[14]

Local unpleasantness did not abate. Tony Palek ordered Os off his land. "You're spreading barkellosis on your feet!" (Presumably he meant brucellosis.) An enterprising farmer who had rigged up ingenious new machines for working marsh peat on hundreds of acres and plowed up the biggest booming ground in the area would not allow Hammy on his land.[15] Longtime aide Dan Berger remembered the revealing statement: "Everything would be all right if they'd only get rid of the Hammerbergs!"—a clear sign that the speaker didn't know either party. Feelings rose high; the department canceled a scheduled February public hearing in the village of Bancroft. Hammy decided that with the program at stake, he would offer to leave to save it.

We could live with this situation, unpleasant as it is . . . [But] we cannot have both a strong program of intensive research and . . . management in this area at the same time under present conditions . . . [and] management [cannot] wait until the local situation improves. Therefore I propose . . . To follow our present schedule of research on the Portage County area only through the spring of 1955, and to move my headquarters away from the area in June. . . . Os Mattson [will] continue to work in the area . . . to further the management plan [and] . . . research on a limited scale. . . . [By so doing] we are very apt to lose the continuity of our carefully related booming ground and banding programs half way short of the full cycle . . . a great loss of research investment—a failure to harvest the crop after much of the work of planting and cultivation has already be done. To us personally, it would be a staggering loss, for this is our second attempt to follow a complete cycle through. It is most unlikely that we will be able to try a third time. . . . To guarantee the management program, research . . . must be drastically curtailed . . . a very great personal sacrifice to all three of us here.[16]

He read this letter to the executive committee of the Wisconsin Conservation League visiting from Milwaukee. [17] They were outraged. Had he had been pushed into making such an offer?[18] They produced a widely circulated declaration of support.

A long period of correspondence and discussion ensued. Hammy produced a "minimal" research approach for the department, listing factors controlling population and ways that management could create them. He and Fran were the only people with the requisite background to conduct the obligatory comparative study.[19]

Kabat, their faithful champion, wrote the Hamerstrom's chief, Bob Smith: "I will, of course, order the Hamerstroms to move to new headquarters if I receive orders to do so from my supervisors." But he went on to insist that the controversy on the marsh was more a matter of intimidation by the opposition than a genuinely felt dislike of the Hamerstroms. "I am convinced that we will have much worse public relations if we, under the new circumstances, order [them] . . . to move to Madison."[20] He answered Smith's objections, pointing out the costs of a Hamerstrom departure and the extra salary for another game manager in the area all winter and into July—to say nothing of the loss of valuable data. Os, of course, would have to stay.[21] Smith temporized. When he asked Hammy to explain the department's intention to area residents, Hammy refused:

I've tried to write . . . but . . . without a definite land policy, I can't write a letter that means anything. I still believe that, in addition to whatever general grassland improvement there may be . . . the scatter-pattern of lands specifically for chickens is essential; and that . . . [it] should be bought by the State or by private individuals (or both), rather than leased from farmers. . . . But I have no authority to say it's going to be done that way—that's a high-level decision of management's. . . .

It will be difficult enough to write the sort of letter that's needed even after the decision has been made. . . . [It is] a public relations job, and a pretty delicate one at that. Without that decision, the letter hasn't got a Chinaman's chance. I'd rather risk your displeasure over failure to write . . . than over the blowup that I am sure would follow an incomplete . . . explanation of land policy.[22]

The matter hung undecided for months, generating much uneasiness. Dave Duffey, a reporter for the Milwaukee Sentinel, reported that the Hamerstroms were to be moved. Their chief [Smith] had admitted that they "were a public relations problem," and the conservation commissioner himself spoke of taking them "out just until this thing eases over." But Duffey went too far; he included a story of the arrest of a visitor at the Hamerstrom home for shooting a grouse out of season, and dragged in a purported "hushing up" of an unrelated case at Horicon marsh as " . . . factors involved in the transfer."[23] The Hamerstroms sued Duffey for libel and damages; he settled out of court.

A new survey was undertaken. Interviews of nearly two hundred landowners showed forty-six persons (26 percent) for allowing private acquisition of land; sixty-seven (36 percent) against, and sixty-two (38 percent) neutral. Seventeen respondents (10 percent) recorded feelings against personnel. Hammy, at a public meeting, heard these figures and jotted them on an envelope. "Personnel"—that would be Fran and himself. Only 10 percent! That wasn't too bad![24]

All the while, individuals like Wallace Grange, Owen Gromme of wildlife art fame, and conservation commission Art MacArthur, a classmate at the Game Conservation Institute, were speaking out.[25] National recognition grew. Hammy spoke at the December National Wildlife Federation meeting in St. Louis The scatter-pattern plan, he declared, "will do the job without the necessity of buying up the whole area. It will guarantee . . . nesting-rearing areas without damaging the local economy . . . the only reasonable solution for many wildlife problems."[26] The Federation voted their support. They hoped that in the fourteen states where farming pressures had destroyed habitat, scatter-pattern management would now preserve pinnates.[27] News releases featuring the recommendations of the "noted man-wife biologist team," were widely picked up.

A Missouri lawyer agreed to deal directly with still somewhat active Peppard and Rudy-Patrick seed companies to get statements of support. (Hammy's hands were tied here. As he wrote, "And—most inconvenient—Vilas Waterman [one of these vociferous characters] is the local foreman for Peppard.")[28]

Finally, in August, the conservation commission acted. Harry Isherwood, too busy farming to go, sent his wife to Milwaukee, where she testified, in all sincerity, that state ownership would cause tax loss, damage to the ditches, and expenses to the district. She was in the minority. The testimony of supporters and friends turned the tide. A longtime supporter, retired motorman Frank Ingalls said, "If you do not keep the Hamerstroms on the job, you are not treating us people right." Harry Chamberlain and Tip Booth, the respected president of the village of Hancock, boosted the program. Division head Bob Smith insisted that research would continue and that Hamerstroms "were not hired by the state to deal with the problems they are now faced with." Joe Hickey, speaking for the Department of Wildlife Management at the University of Wisconsin, called the project "this crucial effort." Supporting letters were persuasive, and the commission approved the project.[29]

Details to be settled remained; they waited until October to hear, "All indications are that research will be continued, but that management efforts may be limited in one form or another."[30] The Hamerstroms would be moved only when the management plan, honed by ongoing research, had been implemented. Private purchase and land leases would continue. The state guaranteed no condemnation, purchase only with approval, and—most important—a ninety-nine-year lease plan, with payments reverting to local authorities to protect the tax base. That crucial provision quieted local fears of losing needed revenue.

Now friends and neighbors drove up to the big house with pleasant news. Neighbor Harry Chamberlain drove all over the marsh singing their praises. "These people wouldn't tell a lie—I know them—and they just love the prairie chickens." And when, according to Fran, Hammy was offered a job in Washington at "three times his salary, he just laughed."

The many residents of the county who saw the Hamerstroms as good citizens breathed a collective sigh of relief. Members of the Grange, that rural organization designed to support farm and rural folk, sponsored a pleasingly timed twenty-fifth wedding anniversary party in June 1956. It featured a huge cake and much goodwill. A very pleasing recognition came in 1957, when the Wildlife Society chose their bulletin, *A Guide to Prairie Chicken Management,* as the most "distinguished and original contribution" of the year.

Gossip and skirmishes continued. Wallace Grange wrote them an irate letter back in 1955 on the basis of an erroneous report that they were supporting an open season on the marsh.[31] A management committee set up by the department, ostensibly to encourage local support, failed. Its die-hard members postponed any consideration of food patches and voted down a suggested program for wider viewing of the booming. "I cannot see," said an independent consultant, "how anything constructive can be accomplished. . . . I do not feel that the committee acted in good faith."[32]

As late as 1981, a small group of violent objectors still felt the same way. Town chairmen in Pine Grove, Carson, and Milladore repeated familiar complaints. "Can't understand why a grant should be given when license fees paid by hunters in the state was [sic] intended for the same purpose." One called the project "Absolute nonsense. This property is apparently a burden with DNR financially, and could be self supporting with other uses. . . . It is a waste of taxpayers monies."[33]

But most people accepted the project and its leaders. "The Hamerstroms?" said a long-time resident in the 1970s. "People don't talk much about them any more." The decisive factors were the public and the media that informs it. With diligence, showmanship, and help from conservationists statewide, the Hamerstroms had prevailed. Hammy knew just what had to happen next and stated it elegantly, in words that are echoed to this day:

We cannot control whatever it is that causes cyclic ups and downs. Neither can we regulate rain and snowfall. . . . Agricultural pressure will continue. . . . All of these factors have been cutting the prairie chicken population down. One thing can be done to counterbalance these negative factors: set aside grassland reserves. Given a place to live, chickens will come through this low . . . and the next and the next—as they have done so many times in the past. This means grassland . . . when the quality of the existing grassland is markedly reduced by continued drought. But if . . . reserves are not set aside now, while it is still possible, there will come a time when just such a combination of circumstances as we are seeing now . . . will prove too much and our chickens will disappear forever.[34]

The conflict had taxed Fran. Reading this chapter, she said, "I can remember dreading getting back here. I never knew what fracas I would find. Once I actually got lost as I drove home from Wautoma, probably because I really didn't want to get home." Hammy put the matter behind him. He sorted out a few important documents—the petition to the governor, a letter or two—filed them in a folder labeled "Controversy," and deposited them out of his office, in the small inactive file in Mary Mattson's old kitchen. He intended to make the most of the time ahead, a time, according to his son-in-law, when he fine-tuned the scatter-pattern, worked with supporters of all kinds to make judicious purchases, and saw skillful, effective management increasing wildlife numbers. These, the most fulfilled of his years in Plainfield, validated his plan and his hard work. He knew what the turmoil had cost Fran and valued her loyalty more than ever.

# 13

# Hamerstroms' Kingdom

## The Complexities of Success

*The quest must be so difficult, and promise such long and devious paths, as to hold out no assurance of ultimate success.*
                                                    —Aldo Leopold

Hammy's reputation was by now secure. His early papers were called classic; his reputation among colleagues was secure, and a growing recognition came from academics. One later summed it up:

Hamerstrom has been the epitome of the crack game biologist. His research has been spread over ring-necked pheasants, bobwhite quail, gray partridge, white-tailed deer, muskrats, horned owls, sandhill cranes, sharp-tailed grouse, snowy owls and short-eared owls. Its main thrust was on Wisconsin's prairie chickens. . . . [His] technical publications number over 50. Two of these . . . involved a scattering of permanent grassland brood-rearing areas in the breeding range of this species. The state in 1957 was not at all ready to purchase these areas and take them off the tax rolls. With Hamerstrom's encouragement, the free-enterprise system stepped in, and at a cost of hundreds of thousands of dollars has now set aside over thirteen thousand acres for prairie chicken management. As the direct result of this magnificent and unparalleled effort, there are now about four thousand chickens in Wisconsin. . . . their highest numbers since 1950–1951.

Hamerstrom's keen critical mind has long been utilized by editors of *The Journal of Wildlife Management.* . . . The man is a super scholar who scorned . . . academia to save a game species that others were ready to write off.[1]

Fran too had achieved some renown. She was seen as something of

194

a phenomenon, but her publications and contributions to the field were beginning to be noticed. Their daily life, however, remained simple and consistent. The crack biologist and superscholar got up first, tended the stoves, made the coffee, and listened to the early morning weather roundup on WLBL, Auburndale. The regular cranberry bog report there advised him of conditions on the marsh. Hammy always dressed to go outside, even to the outhouse: socks, shoes, trousers, shirt, jacket, and—in winter—his long heavy scarf.

After breakfast, he retired to the office, emerging at noon when Fran called for lunch. Their daughter remembers, "Meals were often silent. We all read at meals. They didn't have any other time for their *New Yorkers*."

Spring, as Hammy allowed, was demanding. He wrote, "It is now midnight, lacking only seven minutes. Please excuse delay in thanking you for the letters and slides. Spring this year was the latest ever, and our wonderful society unexpectedly produced seventy-five guests to fit into the booming schedule. . . . I've just been so dead for sleep for so long that I let a lot of important things wait. My apologies."[2]

Fran managed a varied schedule with her customary energy and flair. Surprisingly, once the pattern of booming was routine, she did research of her own: "It had to be low profile," she said, "but I had always been interested in birds of prey. Hammy and I had published a study of what young raptors ate when we lived in Michigan. Now I needed a project of my own. After so many years as a wildlife biologist, I tired of having people ask me if I was interested in birds too!"[3]

Two chapters in *My Double Life*, "My First Falcon" and "Will She Die without Rangle," describe her stubborn, secret, and ingenious beginnings as a falconer. At the public library she found a rare book, and in the face of a suspicious librarian, deciphered the old English enough to find key words: austringer, jess, creance, musket, and rangle.[4] About her falcon she said, "For part of my life, she was my whole world," and she managed that world in her own way. It is customary to carry the falcon reins in the left hand; she always used the right. She found nests in Iowa; and in Necedah conducted the hawk census, counting seven species.

Now she returned to that passion. Even during the booming, she could squeeze her study into free moments between the thirty hours each week of departmental assignments and her role as chief cook. In booming weeks, that meant producing hearty breakfasts and second breakfasts for observers, plus fixing lunch and supper for the family, crew, or the unexpected guest. "Hammy got used to my inviting strangers for lunch." She started keeping hawks, and the shed became her mews.

This fit their principle: to spend one-third of available time on "exactly what you please." They always took advantage of the few easy weeks of fall—ideal for study vacations. In 1956, they explored part of the original chicken range. Missouri conservationists provided maps of their census routes; the Hamerstroms packed the Volkswagen and headed south via Effingham, Illinois. There was a good dance band there where they danced well into the night. In Missouri, they took country roads to Jefferson City and south to the Chmelir Farm and Prairie Camp. Hammy looked it over: "By that rise is the best spot for the tent." As they carried the gear over, they flushed a flock of 43 chickens. "Get it in the journal, Fran. Here's the map: NE SW 6, T43N R21W should do it." Next morning they scouted census routes, stopping often. "Look, Hammy!" crowed a sharp-eyed Fran, "Feathers! And droppings!" Often they found chicken populations down or vanished.

Hammy's report to the DNR the next winter made the point that people in Missouri, as well as those in Wisconsin, were likely to blame diminished population on state wildlife personnel rather than on habitat loss. Two men repairing a bridge reported that few chickens were left. They blamed "government men who had been in a few years ago and trapped the birds out." Of course, trapping and releasing a few chickens could have had only minimal effect on numbers.[5]

They admired the big bluestem grass, six feet tall. The owner of a nearby farm told them that some of that tall-grass prairie had never been plowed. Two leisurely days later, they drove on, through the Kansas Flint Hills to Nebraska and then Denver. There they had a lively reunion with friends and fellow biologists at the meeting of the American Ornithologists Union (AOU).

Back in 1952, they had been two of the thirteen people who met to discuss prairie chickens at the 1952 National Wildlife Federation meeting. That small gathering evolved into the Prairie Grouse

Technical Council (PGTC), whose 1960 meeting would be held in Stevens Point. Hammy spoke; his presence, reputation, and calm, resonant voice commanded attention as he traced the history of the chickens in areas they had explored and related it to his recommendations for improving the range. "You could have heard a pin drop," said Ray Anderson, then working on a Ph.D. with Hammy as thesis advisor, and who later became a professor at the College of Natural Resources in Stevens Point. Hammy stressed chicken responses to accidental environmental changes as important signs that response to deliberate changes was an achievable goal.

It is characteristic of Fran that when I asked her about the PGTC she said airily, "Oh, that? That's just something Hammy and I did on the side." Not so: it was an effective group. The organization's records reveal that in 1950 chickens were gone in Iowa, almost so in Ontario and Indiana, and precariously surviving in Illinois, Michigan, and Wisconsin—except on the Buena Vista.[6] Missouri had remnants; Indiana had only one refuge. Management consisted mainly of manipulating the hunting season plus a little land clearing, and feeding in winter. The new forum allowed research to be shared; even related prairie chicken—Attwater's in Texas, and the lesser prairie chicken in other southwestern and mountain states—were included in the attention.

*A Guide to Prairie Chicken Management* had defined the keys to management. Once principles were clear and geographic variables (in preferences in food and cover, the habits of scattered flocks, and suitable cover in the nest brood period) determined, suitable programs began to be put in place. The reports of the many observers of the booming on the Buena Vista helped generate the necessary funds for buying Wisconsin land. Other localities might require other approaches.

Hammy was, in effect, the glue that held the group together, writing hundreds of letters and memos and attending meetings with Fran, who provided both chemistry and support. His letters to men facing many of the same problems that he did reveal an attractive, often funny, sometimes testy, realistic, and lovable man. Don Christisen, a long-time member of the organization, says, "If the PGTC had a father, it was Fred Hamerstrom. (Of course, we all knew who the mother was, Fran.)"[7] He shepherded the scattered association through its transition to a functional, independent organization. "There was a meeting of prairie chicken technicians in 1961, in Pierre, South Dakota. A field trip afternoon was canceled . . . [so they] sat around a small hotel room and groused about the lack of direction [from the National Wildlife Federation]. It was then that Fred's leadership outlined what should be done. By the time the next meeting was held . . . the group had formed the Prairie Grouse Technical Council and shed the sponsorship of the Federation."[8]

Results were gratifying: by 1964, thirteen years after the first small caucus, one hundred people, including managers from Illinois, Michigan, Missouri, and Wisconsin, attended council meetings. They brought sophisticated papers, told of land purchases and attempts to transplant flocks. Their publication, *The Prairie Grouse News,* was widely distributed. The membership, replete with men whose primary interest in conservation had been preserving huntable populations, evolved into champions of bio-diversity and grassland protection.

Hammy's humor muted contention, as one young chairman vividly remembers. Ron Westemeier, Hammy's former thesis advisee, described the 1967 gathering. New, nervous, and with his boss in the audience, he was worried; he had to announce that the expected dove hunt had been cancelled. Professors, game managers, and forest service employees, along with businessmen and members of the Izaak Walton League or the Audubon Society, some with shotguns at their sides, came from South Dakota, Nebraska, Kansas, Oklahoma, Missouri, Wisconsin, Texas, Ohio; Winnipeg and Victoria.

Westemeier stood, welcomed the crowd, and then hesitated. He cleared his throat. "I'm sorry to tell you that we won't have our usual dove hunt this year." He raised his voice over the ensuing buzz. "Our governing board has decided that hunting, for a group of ecologists, would be a perversion of conservation principles." Frowns and audible grumbles greeted the statement. Then a deliberate, resonant voice rose above the buzz.

"Now, that's my kind of perversion!" Laughter broke the tension. Hammy, smiling, settled back in his chair, and a productive meeting ensued.

Hammy served on every kind of committee, his authority lightened with humor, his standards tempered by realism. "The question of proceedings is a sticker. I've hollered louder than anyone else,

I fear, that an informal no-nose-buried-in-prepared-paper presentation is what gives our meetings a special flavor. . . . So what have we got? No proceedings. Maybe I'd better pipe down."[9] Later, after four years on a bibliography committee he saw its futility. To consolidate over two thousand bibliography items, some on three-by-five cards, others on punch cards, and to trace original sources: "scattered from hell to breakfast, and there is no central place (short of the Library of Congress, perhaps) . . . where one could settle in and go through the whole bunch," was simply too much. Besides, he said: "Keith Evans has published a fine bib. Does committee need to go on?"[10]

The scope of Wisconsin research enlarged after a 1960 National Science Foundation (NSF) grant brought support for two graduate students to work with them.[11] Westemeier produced a history of land-use and cover types in western Portage County, and Ray Anderson concentrated on the booming and confirmed experimentally much that they had suspected. Pheasants do dominate booming grounds and chase the hens off; prairie chickens dislike trees; cocks attend "their" territories from mid-September through early January; and individual males return to specific booming spaces from spring to fall to the following spring. They can be stimulated to booming activity with broadcast tape recordings of the sound.[12]

As they bought land, workloads enlarged. Hammy, who once expected to hire few paid workers, now added seasonal helpers from January until June. A spring regular was close friend Dan Berger whose avocation was—and is—to man the Cedar Grove hawk trapping station on the Lake Michigan shore during migration. When hawks were few, he drove through the winter-brown hills of the moraine to Plainfield. "More and more," according to Dan, "Hammy had to be content to supervise the fieldwork, except at booming, when he always was in the field. Fran and I did most of the trapping, banding, and weighing." Occasionally Dan did the observer debriefing sessions. "I didn't like to. I was in awe of Hammy, who expected a lot of himself, and of others. My debriefing often seemed lacking. If I reported a successful copulation, he would ask, 'Didn't you ask about the hen's behavior?' Naturally, I hadn't. And he always rechecked my notes. That sent a certain message."[13]

Growth had its price. The fallow acres created by the Soil Bank in 1960 meant more revision of maps and longer reports.[14] Then, in 1961, Wallace Grange retired from his 9,500-acre Sand Hill Game Farm, and the state acquired it.[15] Hammy scrawled a memo: "Grapevine says Os will be in charge of Grange tract and continue PC mgt. here. Congratulations to you both—advancement for Os and a damn good man in charge of the new area . . . a very important new operation to my way of thinking."[16]

An upbeat 1962 report showed a year's research budget of $7,700 a year—about half of the 1958 expenditures. Those dollars supported 4,554 acres of grassland under long-term lease. On one two-hundred-acre tract, mowed and sprayed to eliminate willow two years before, there were now "2 chicken broods, 1 upland plover nest, 3 does, 3 fawns, and a new booming ground with 5 cocks! . . . Its food patch fed a flock of some 75 chickens and 15 quail last winter! . . . Wildlife does respond to land treatment!"[17]

Management was demanding: nest-brood cover required permanent or late-mowed grassland, not fields grazed billiard table clean by cattle. Food for winter and the critical brood-rearing period was ensured by a careful planting sequence. Standing corn followed by oats seeded with timothy and alsike clover provided food the first year, autumn food the second, and six years of brood and nest cover. Of course, each of thirty-three managed units might require prior burning, spraying, cutting, or even bulldozing the walls of aspen that sprang up like dragon's teeth.[18] Os supervised all such operations.

He wrote the management reports and gave papers, too. The "Os look," skeptical, over the top of the eyeglasses, was often seen, and words augmented the look. Berger described the time when Hammy told Os about an observer's report that a certain cock was displaying, adding, "At least the observer thinks so, though he wasn't quite sure." Unexpectedly, Os responded, "Oh well, you tell them what they're going to see, and they see it." Hammy stiffened. "That's a damn poor thing to say to a man!" he roared. But he got the point. Did his briefings program observers? He dispatched a skilled aide who found that the cock in question was indeed active at that booming ground. In the early years of sharing the big house, frequent conferences were routine. After Os and Mary moved to their own conveniently located house on

Paul Olson (in wheelchair) in the late 1960s, with an unidentified friend, Fran, and Hammy.

the marsh in 1962, less frequent communication allowed differences more easily to arise.

Hammy had expected substantial changes in their first plan. "You just can't get all those small parcels," he said. He wanted well-distributed parcels, separate from existing large blocks.[19] In a 1963 paper to the Prairie Grouse Technical Council, Os made his own recommendations. He called for a "minimum size [of] . . . two hundred acres . . . one two-hundred-acre managed grassland unit in the vicinity of an established booming ground" to insure that prairie chickens would remain during population lows.[20]

Hammy hadn't seen the paper. He wrote Jim Hale, "taking objection" (he crossed out the word and wrote in *exception*) to the paper's premise: "Experience has now indicated that fewer but larger parcels will give more assurance of survival." That statement, he felt, was too strong. He added, stiffly, "It would seem to me more appropriate for Management to have discussed such figures with Research before making them public." He knew that if he said nothing at the meeting, he would seem to agree with Os, but disagreement might "jeopardize the land purchase program." His postscript softened the message: "Most of Os' paper is very good, as you will appreciate. . . . The main thing is to get the job done . . . and that is going forward very well & Os is a key figure in the operation."[21]

As indeed he was. Jim Hale's tactful solution was to suggest that since neither size of reserve had been tested, claiming that one size was more effective than the other, might create confusion. Could Os back up his point with proof?[22] Increasing chicken population clearly involved many factors and many variables. Specifying size precisely seemed unwise.

Change continued. It had to after Hammy froze his toes. Fran explained, "A hard winter came in 1962. Someone tightened the bindings on his snowshoes; Hammy didn't adjust them: he didn't expect to be out long. When he came in his toes looked exactly like expensive purple grapes. Dr. Garrison feared amputation, and insisted on winters in a warmer climate. That's when we started going to Texas and Mexico."

They saw that they needed to accomplish fieldwork with less manpower; traplines took too much time. From the very first, they had seen hungry chickens congregate where there was food in the winter; now they experimented with a technique developed by a turkey farmer they had come to know. Bait and a forty-foot fishnet fitted for quick release over a promising area worked for them, too. Once rigged, it was only a matter of timing, replenishing the bait, and keeping the cotter keys greased.[23]

In spite of all delegation and simplification, Hammy never had enough time. Supporting the efforts of the very active Tympanuchus Society, the Prairie Chicken Foundation, and the other organizations in which they were active members, had to continue. There were "Too many fires to put out!" They ranged from hurt feelings when credit for land purchase was not given, to requests for low phosphorus marsh soil for a soil scientist's experiments.[24] Requests for reprints had to be filled.

They continued helping—and hosting—researchers from Europe; Hammy wrote hopeful students about conservation careers and crafted many effective letters of recommendation. In 1963, as the national peregrine situation became desperate, Joe Hickey asked him for the name of someone who could run a peregrine census in the East. He did not hesitate. Dan Berger was the man. "I'd say he should be [in] complete charge even if he has a Ph.D. as assistant, and he should choose . . . his assistant. If Dan and I were to do this, I'd expect to be the assistant—and I don't think you have any potential assistants . . . [with] better qualifications than mine. Keep us posted, hey?"[25] Dan did the census, helped by Helmut Mueller; Fran arranged that the two men would stay, while in Massachusetts, with the ever-cooperative Posey.

They kept up with grouse elsewhere, even at the cost of hasty returns to the deer season, the sudden advent of winter trapping, and the piles of paper and data on his desk. They attended most International Ornithological Congresses (IOC). They were in Basel, Switzerland, in 1954, and in Scandinavia during a "wonderful" three-month trip to Europe in 1958.[26] In 1962, they, with Fran's horned owl, attended the IOC at Cornell. In October 1963, they made the long drive to the Northwest Territories to scout for sharp-tail. In 1964, on a hunt with Farrell Copelin, in Oklahoma, they thrilled to the gobbling of lesser prairie chickens—"more like rookooing than sharp-tail gobbling, even. Fantastic!"[27]

They fostered burning conservation issues. Fran recalled a meeting at their home, when some of the fiery advocates of a legislative ban of DDT in Wisconsin considered strategy. After a good deal of work and considerable input by the Citizen's Natural Resource Association, and thanks to an existing Wisconsin law forbidding the addition to rivers of anything that could harm the fish, the ban passed a full year before the national prohibition.

In 1970 Hammy, alerted to problems by Thomas Whiteside's article in the *New Yorker,* and the record in Vietnam—"a shocking, disgraceful story"—wrote to Jim Hale, urging departmental investigation of 2,4-D, the infamous Agent Orange.[28]

The department encouraged his giving papers and publishing. Even with Fran, Os, Dan Berger, and others, he could still deal with only part of the growing mass of data on prairie chickens. Matters of real interest had to be ignored, such as the decline of specific booming grounds as farm practices altered, or the remarkable tenacity of the prairie chickens that persisted in booming for fifteen years in cultivated fields that now covered their ancestral booming grounds.[29]

He knew his limits. He refused a departmental request that he develop "economic means to maintain habitat"; that was a function of management, not of research. Friends and admirers suggested wider initiatives: "I'm no damn good when it comes to . . . popular writing and influencing legislation . . . someone who knows the ropes in these two highly specialized operations is needed . . . I'd be . . . worthless as a prime mover."[30]

Yet his letters to colleagues reveal growing themes of overwork and uncertain support, inadequate funding, and insufficient time to educate and publish. Where there had been half-jesting complaints: "Enough: let's go fishing. Do you know, I haven't been fishing for three years: That won't do," he now asked to be freed from auxiliary responsibilities when a difficult winter and sickness had rendered him "half effective."[31]

Their NSF grant ended in November 1963. They were told that state funds were uncertain and had to wait well into 1964 for a decision. Even then they were told that support might continue only for two or three years—a blow indeed to ones committed to long-term research. It was ironic: with a cyclical population rise expected, the project was threatened. They had studied falling populations twice, but no one had studied rising ones. Indeed, the entire year—whose high point was Elva's June wedding, with a fine Hamerstrom party on their spacious lawn—was one challenge after another.

The month before those festivities, a late spring memo from Os announced a totally inappropriate tree planting that had just ruined an important booming ground. "I had to stand by and watch as they were being planted. The fellow who had contracted to do [it] . . . said the site was no good for Norway pine. They will grow enough to ruin PC range . . . he is planting them for timber production—not Christmas trees. I am calling this the most damaging planting in the history of the BV project. Agree?"[32]

They faced the continuing—and increasing—apprehension that mechanized farming could undo all they had done. Drought in the sand counties meant that almost every summer there would be six dry weeks. Irrigation, the solution, brought "progress" with it: pesticides, herbicides, economies of scale, larger equipment, higher capitalization, and corporate farms practicing concentrated cropping. We saw it happen: our neighbors (with eighty- to three-hundred-acre small farms) sold or leased to one or another of the operators who could afford to pay a premium for land adjoining theirs. One entrepreneur who spent winters in Florida, reportedly balanced his portfolio with apartment complexes there.

The Hamerstroms had lived with drought, wet summers, late winters and springs, but these new onslaughts of road-to-road cropping, chemicals, and bottom-line bookkeeping were threats from which the chickens might not recover. Wells put in by a new reverse-flow method were quick to drill and inexpensive. An operator drove a shallow sandpoint in the sand and flushed pressurized water into casing sections, added one by one until they reached the required depth—usually not more than 100 feet. In two or three days the farmer would be pumping a thousand gallons a minute from the huge reservoir in the bed of pre-glacial Lake Wisconsin.

After 1964, special electric rates for high-volume users brought trouble-free electric motors. Farmers cut windbreaks and replaced laborious hand-moved pipe with large water-driven irrigation systems. They bought, cleared, and leased land for ever-larger operations. Land values mounted. It remained for the Grouse Project to buy more and more land.

Hammy filed a *Rural Electric Cooperative News* supplement that described these developments in glowing terms. He underlined one sentence: "Geologists and other experts studied this deep natural reservoir of water and concluded that no foreseeable amount of irrigation pumping could significantly deplete it."[33] He was skeptical, as we heard at a joint dinner. "It's all right for someone like you, Kip," he said. "You are running a family farm, and you live on and manage a modest acreage. But these fellows who get thousands of acres! and pump day and night! and cut down all the windbreaks! and clear the woodlots! and spray from airplanes . . ." He gave his head an eloquent shake.

He appreciated the function of chemicals: the necessary clearing acres of brush and willow on the land they bought would have been prohibitive without them. Airplane spraying was another matter. The pre-dawn din of the dipping planes was infuriating, but he deplored the bird mortality from insecticides like Parathion—and when the Hamerstroms discovered that operators were mixing leftover materials and pouring them out in an area with a high water table, they became deeply, and rightly concerned.

He raised the alarm, writing to Jim Hale about the expanding potato farming. "It's the irrigation on muck that has me scared. . . . somebody wants to lower the water level, there is a drag line working today deepening the lateral which runs on the section line between secs 4 and 5 22–8. . . . Looks to me as though we have a first-class emergency just around the corner—or maybe already in our laps. . . . [May] a few irrigators . . . so lower the water table as to interfere with the operations of the non-irrigators (cattlemen and prairie chicken fanciers, for example)?"[34] To top it all, ironically, someone in the Society with a game farm up north lured Os to a better paying job.[35]

Early the next year, Hammy was still "snowed under with chores, have just been riding along on your efforts. Right now I have a portable typewriter balanced on top of a bag of corn while I watch a trapping station: just caught 4 chickens, the first this winter (the old complaint—no snow)."[36] But by June, he promised that committee work would be "possible," in spite of "reports by the yard, letters to answer, and maybe even . . . a look at the data after the chores are done (a hell of a way to have to arrange things, isn't it?)"[37]

It was an uneasy time, for in spite of all efforts, the chicken population remained small. The expected rise after the peak of 1952 did not occur in the 1960s. Numbers dropped to an unsettling low, particularly on Leola marsh, where the decrease was drastic. In 1965, only three certain booming grounds were found on the fifty-thousand-acre Plainfield study area. "Population is critically low, chickens are on the point of disappearance, as they are through most of the Middle West," Hammy wrote the Department. "Michigan, according to Ammann, has about two hundred adults. Indiana flock may now be gone. Westemeier thinks there are less than 250 in Illinois. Portage County is the exception," he concluded.[38] Yet, "It would be impossible to undertake another winter of banding and at the same time hold to the schedule of analysis and writing without extra manpower."[39] And after 1966—the last spring count they would staff and manage—the University of Wisconsin–Stevens Point handled the spring census. In 1969, Fran and Hammy joined in the count on the booming grounds for only three weeks, and dropped that to one week in 1971.

Fortunately, the upturn came; in 1967, numbers increased on part of their managed area, and a satellite area on the Little Eau Pleine showed high counts of prairie chicken, and even higher ones in 1969. Then, finally, in 1970, there was a big (62 percent) increase in the count on the Buena Vista, with a substantial spillover into Leola, in spite of the spring floods there.[40] Hammy decided: never again did he want all the project eggs in the Buena Vista basket. He called for the establishment of new colonies to permit rebuilding should disaster strike there.

Substantial honors came to them but did nothing to change the reality of their work situation. In 1968, the Department downgraded Fran to seasonal status. "Most inappropriate," he wrote Jim Hale. "She should not still be at the Biologist II rating." Her years of training, her service, high quality work and honors, and the many unreported extra hours she had put in, should at least justify a maximum merit award.[41]

They could now reflect, in the quiet of their solitary moments, on their twenty-two years of meticulous research. They had studied long-term populations, fluctuations, density and turnover rates, established the daily and seasonal movements, and the survivorship of chickens. They had related all this to habitat type, quantity, and quality, enabling the creation of working management plans. They had recorded the impact of raptors and predatory mammals on the booming ground chickens. They had compared the mating displays of North American and European grouse.

The prestigious National Wildlife Federation recognized that record with their 1971 Award to the couple. Typically, after writing their thanks, Hammy added, "When I went to pay the bill, I was told that the Federation had signed for it. This may be more than you intended (in addition to the handsome expense checks you gave us), and I would think it appropriate to refund to you the hotel bill for the last three days."[42]

Later that spring, he read an interview with a marsh farmer in the *Stevens Point Journal.* He had, the past winter, fed several one-hundred-pound bags of corn to a flock of forty or fifty chickens. He was pleased at their modest comeback. "I like the birds!" he said.[43]

Hammy showed the report to Fran. "It appears to me," he said with serene satisfaction, "that our plan is working. Perhaps our part of this job is done."

# 14

# A Naturalist Family

*As far as the education of children is concerned . . . they should be taught not the little virtues but the great ones. Not thrift but generosity and an indifference to money; not caution but courage and a contempt for danger; not shrewdness but frankness and a love of truth; not tact but love for one's neighbor and self denial; not a desire for success but a desire to be and to know.*

—Natalia Ginsberg

Meanwhile, there were the children. Although critics were quick to characterize the Hamerstrom approach to child rearing as too casual, the singularity of their parents may have had more effect on them than the actual family style. In a small conformist community, difference—an unpainted house, peculiar occupations, and the fact that the children called their parents by their first names—created a disconnect from school staff and peers.[1] Nonconformity outweighed for many the considerable similarities between them and other families.

However ahead of their time they may have been, the Hamerstroms maintained regular routines and recreation, especially in the frequent periods when they saw few outsiders and interacted as a unit. The four of them skated together at the Lonor, a cheerful rink with a good floor nearby where they took the free dance lessons offered in the 1960s. The Stems recall Elva as almost part of the family. "She said, 'I don't plan to treat my children the way my family treated me.' Then she stopped and thought, 'Though I guess I turned out okay!'"

They spent time with other families—mostly outsiders. They hunted with Bob and Marie McCabe and their boys, and in some summers, visited the Croys, a couple with children who lived in spacious comfort on a lake near Waukesha where the two families relaxed together. Alan stayed with the Croys for two summers of work in Milwaukee.

They maintained family connections as much as possible, with weddings the favored occasions. A Christmas with brother Dave and Liz Hamerstrom, another with Putnam and Dottie Flint, preserved ties. Before he died, Darrow had asked them to "take care of Ruby." Although Hammy had remarked to Os, "Taking care of Aunt Ruby would be beyond my possibility," he took a week off in 1953 to move her, now sadly diminished, from her apartment in a formerly elegant Chicago hotel to a retirement home in Wisconsin.

The children loved Posey, their sole remaining grandparent. At first, Fran would take them to Chicago, where Posey would meet them and ride with them to Boston for a summer with her at the Hull house. Soon they went in a Pullman sleeper by themselves. Elva remembers her grandmother with great fondness but adds, "Going there was just something we did—like brushing our teeth every morning. Given a choice I might have chosen to stay home with my pets [as she did before too long], my bike, my freedom and the old irrigation hole."

They augmented their family, trading their children and hosting the replacement European children between 1952 and 1955—the height of the prairie chicken controversy. Fran wanted her children to have some of the formative experience she had had; the European parents hoped to build and foster independence in their young as well as inoculating them against shallow nationalism.

Some thought that Fran favored Alan. She called him "my man child." One friend said, mildly, "I thought it was a bit much, but she always was excessive. I remember when she stopped the car and washed him off in a roadside ditch after Mrs. Leopold asked her if she didn't bathe him as she changed his diaper." Perhaps she did indulge Alan. He had a dog in spite of her aversion to dogs; a spacious room of his own, an electric train, and a gun. But she was also permissive with the European children. Alan testifies that he did pretty much what he wanted—not surprising with a father whose own upbringing was liberal and a mother reacting against the constraints of her youth. His mature judgment is that his mother's self-confidence peaked at the time that the European children visited. It was the period in which she proved to herself that she had moved beyond her own upbringing and could deal with young people in a sensitive, respectful, and productive way—a mode that prepared her for the role she would play in the life of the gabboons.

Alan Hamerstrom had great freedom of choice in a democratic household with practical conventions and parents who paid as much attention to joint and individual pleasures as possible—fishing, some hunting, and family outings. The wood stoves, the unpainted siding, the outhouse, and the hand pump made sense. "Others, particularly school children, were really turned off; I was on a very low rung on the social ladder at Plainfield, especially during the Prairie Chicken Wars." He went on to describe family conventions:

"We had nothing to do with Mother's Day. Fran was adamant—it was commercial—and Hammy agreed in a quiet way." Making do with money was no hardship: "We did without some of the 'necessities' in order to enjoy luxuries like travel. It wasn't till years after Posey died in 1970 that they started to loosen up and stay in a La Quinta motel instead of a Thrift Inn. They wondered why I didn't rebel as teenagers were supposed to do. But what was I to rebel against? My parents respected our decisions and encouraged us to think things through for ourselves."

There was little fatherly advice, and no pressure towards a specified career. Should Alan speak of research, he was encouraged, as he also was if he mentioned a psychology degree. When he left Harvard to continue at the University of Wisconsin, Hammy was supportive; he applauded Alan's solo canoe trip from Minneapolis to New Orleans. Alan repeated his previous question, "How could anyone rebel against that kind of upbringing?" Hammy had, however, the final word:

He trained us thoroughly and persistently about grammar and manners, but he was permissive in the big things—except one. He and Fran saw absolutely no basis for believing in a Deity, benevolent or otherwise, and described religion as a tremendous hoax foisted on the masses—an effective way of getting people to do what you want them to do. They did not want us to be manipulated and controlled in that way. Sometimes I felt they were shielding me from religion so vigorously that there must be something I should find out for myself. I remember going to a revival meeting in high school with Elva and a couple of other kids. Elva almost responded at the invitation to come forward and "give your heart to Jesus." She got up—but I pulled her back down.

Young Alan and the "Jackrabbit," circa 1952.

Everyone in the household respected Hammy's authority. When Elva misdialed the new radio that he counted on for the weather report his displeasure was obvious. She never touched that radio again. Elva had been, according to Dick Hunt, an infant who cried a great deal, but she became a pragmatic, undemanding second child. "Hammy was always there for me," she maintains, but "I thought that if something happened to me, it wouldn't matter much. My parents would just go on with their lives."

Elva's life was full. She loved visiting the Mattsons' apartment. She played dress up, savored Mary's cookies, and played with Sarge, their arthritic old dog upon whose back her pet rooster would sometimes climb. If Sarge seemed especially creaky, she and Mary gave him aspirin, carefully, so he wouldn't spit it out. She cared for her own pets: Minerva (a great horned owl), a ferret, and a black rabbit, as well as helping with others—Bachhus Western Union, a screech owl who had been caught by a cat and mauled, for example. Talented with animals, Elva at one time planned to be a vet.

In 1951 when Fran and Hammy visited Europe, they arranged for

Elva at the age of eight with "Hoot Mon," a young great horned owl, April 1955.

Alan to attend a western camp. Elva didn't want to leave her pets. She went up to the neighbors at night and spent the days at home. Her parents asked the sheriff to check their property occasionally, and when a deputy did so that summer, he heard her playing her clarinet in the back garden. He found her. "Where are your parents?" She answered calmly: "In Finland, I think." In the 1980s, a social worker told me—with a frown—that they had heard about the incident but that "inquiries prevented an investigation."

Elva was docile, but she could draw the line. She hated that Alan always went first in the weekly tub bath, not from a principle of fairness but because she had to reuse the scummy water. She solved that problem as soon as she could: she simply showered at the high school gym. She opened her window as required every night, unless it was 20 degrees below. One night she had twenty blankets on her bed. "I wondered why my bed couldn't have been heated by a wrapped rock—or any concession to comfort. I supposed that if I said anything, my parents would say, 'That's just the way we do it, Elva.'" Fran would not buy her a bra, even though she looked, at twelve, like a mature college student. "But when Gretl Lorenz met me at the steamer [on which she had sailed from New York], she took me right from the dock to a store to take care of that."

Elva helped with fieldwork as school permitted. She most often accompanied Fran on the trapline. ("It was a real treat to go with Hammy.") The sense of responsibility enlivened her; the excitement of anticipating the catch lasted through the twice-a-day, sometimes three-hour chore. She prized their picnics at Keystone Rock and evenings at Lake Huron. "After supper in those long summer evenings we would go up to the lake. Hammy, in his cherished rowboat, fly-fished from the prow while Fran and I used hand lines. Sometimes our lines caught twenty-five or thirty small fish to his one—which might weigh as much as all of ours. We always cleaned and cooked them—I still love blue gills. In later years, those excursions became fewer."

Her single friend was Nancy Jorgeson, a neighbor and after-school playmate—they did not hang out at school. At eleven years of age the two girls bicycled all the way to Lake Michigan. They took hilly back roads the sixty miles to Neenah. "It almost did us in," but after a night there with the friends Fran had alerted they went on to Two Rivers for five days of vacation with one of Fran's friends from booming times. After a look at the bikes, she loaded girls and bikes into her car and took them within twenty-five miles of Plainfield before sending them on their way. After a flat tire near the village of Wild Rose they walked to a large old farmhouse where they were obviously unwelcome: "We wished we hadn't gone in when we saw people strapped to chairs in the dismal room with peeling wallpaper where the phone was. We were glad when Fran answered, and left hastily as soon as we had told her exactly where we were." They thought it was a "lunatic asylum," but it was only a poorly run old folks' home. "I felt I wasn't adventurous enough to please my Mother, but I believe that bike trip did impress her."

The accounts of the European children are captivating. The exchange came about during Gustav and Neni Kramer's 1952 visit. "As we sat at this table," Fran said, "drinking red wine, someone remarked that it would be fun to exchange children. We laughed—but within twenty minutes we were talking about which child should go first. Kramer's eldest, Peter, was the obvious choice, but Neni thought he wasn't old enough. So I said, 'We'll send Alan.' He was twelve. Later, on a visit to Konrad and Gretl [Lorenz], we arranged to bring Dagmar over with Peter."

Alan stayed with the Kramers in Germany from November 1953 until the next May, attending a German school as a guest student. He did not speak any German, and he had a lonely time. The next fall, Peter Kramer and Dagmar Lorenz came to Plainfield; Peter for a year; Dagmar a semester. (She was too diffident to say, in January, that she would really like to stay through the spring.) Next Lorenz Kramer and Elva changed places. Elva stayed with the Kramers from November 1955 until April 1956, when she went to the Lorenz home in Austria until August. "I felt like a duck out of water."

The last of the Europeans came in 1958. That May Fran and Hammy had sailed on the Gripsholm to attend the International Ornithological Congress, held in Helsinki.[2] They visited a Professor Lars v. Haartman, the president of the International Ornithological Committee of the Ornithological Section of the International Union of Biological Sciences, who lived at the family estate in Lemsjoholm. Fran was delighted: "We stayed with a Baron!" A subsequent field trip to Lapland took them all the way to the small willows and scrub birch above the Arctic Circle by plane, bus, motor- and rowboat, horse, coastal steamer, bicycle, and foot.

They followed ten tents of Lapps herding some ten thousand reindeer, heard birds sing at midnight, and observed Lapland ptarmigan and fast-maturing willow grouse, species that were well adapted to the brief summer in those high latitudes. In Swedish Lapland, they admired the controlled burns that permit pines to flourish and the management that resulted in abundant hazel grouse, black grouse, and capercaillie. Fran wished that they could share the Lapp diet of reindeer meat and milk but had to be content with the European diet of bread, potatoes, and what she termed "almost inedible" salt fish served in their bed-and-breakfast.

After visiting the Max Planck Institute in Germany and Konrad Lorenz in Austria they brought eleven-year-old Elizabeth Kramer and seventeen-year-old Silvia Ebner-Hain home with them.

Interviews with four of these five Hamerstrom "adoptees" in 1991 highlighted their common memory of freedom, both from the rigid demands of European schools, and in the Hamerstrom home, from parental expectations.[3] Dagmar Lorenz was placed in Elva's grade school class. She came with "quite good English." Her teacher asked, "In Austria, do you have electricity?"

"I knew immediately how seriously I must take that school," she said with a smile, "even without the stories repeated there about the Hamerstrom house—which I put in the same class as a pamphlet someone gave me that called Frank Sinatra a *communist*."

For life with the Hamerstroms was simply enchanting. Fran loved to have children about; there was never any sense of strain. Hammy, a quiet, reserved presence, was tolerant, and very polite to all of them. Dagmar saw frequent squabbles between Alan and Elva that grew louder and louder, until a deep grumble from the office quieted them. And when Fran and Hammy had to be away for a few days, the young people were left alone in the house—another unexpected freedom.

The pump in the back entry delighted Dagmar; she loved lifting and pushing the handle, and the sudden gush of cold, pure water. Accustomed to black bread and liverwurst, she remembers the white bread, "angel food bread," the waffles they made in an electric waffle iron, and the baked potatoes. Kool-Aid was a favorite drink.

"We all had certain chores. We fed the chicken and the pigeons in the barn. Two of us, in turns, washed the dishes and cleaned up the kitchen. In the evenings, we played cards—canasta, I think it was called. Hammy played with us; he seemed to enjoy it. We went on a boat trip to Wisconsin Dells, and on picnics in the canoes. We clambered among those upstanding rocks—Lone Rock, wasn't it, and Keystone? Once or twice we sang in the car. 'Please sing, Hammy,' Fran would beg. Once she whispered to me, 'He sings beautifully.'"

On one unforgettable winter day they built a snow fort in the back yard and slept in it. In the morning, they opened their eyes to a brilliant blue sky. They hopped, still in their sleeping bags, through the crystalline whiteness to the outhouse. For Christmas they cut a tree, upon which Fran clipped old-fashioned candles and hung the costume jewelry she took from a box onto the branches.

Before hunting season Fran talked to her. "Should we get you a license, Dagmar? You don't seem very interested." Dagmar's father did not hunt, and his daughter was secretly afraid of guns. Recognizing that an interest in Nature was imperative in that household, Dagmar announced an interest in plants. "How relieved Fran was to find that I really loved flowers! She encouraged my botanical interests."

Dagmar saw another side of Fran when Posey came. "Posey took up a broom and started sweeping the living room. Fran spoke shortly, almost cruelly to her. It seemed that everything was a provocation. That archaic theme—the adult sense of the mother as threatening to a child's self-worth—was played out to the full."[4]

The couple made a lasting impression on the sensitive twelve-year-old. She sensed a mystery about them; she felt the bond and envied the aura. "I never heard one word of disagreement. There was no sarcasm, little didactic talk at the children." These were clearly aristocrats, not only in the symbolic details—the heavy sterling, the beautiful table and chairs, the sense of style—but also in Hammy's bearing and his impeccable manners. Fran told them the story of their romance and of her trip to California. "In my imagination, I filled it in. He wrote her, or telegraphed, 'Come back! It's me or Hollywood!'" She noted the way they knew and accepted each other's frailties and cherished each other. When we spoke of his death, she wiped her eyes. "It is so touching," she said. Her adult assessment was precise and perceptive: "Their life

Dagmar, Karl Lorenz's daughter, in Europe, 1956. Photograph by Elva H. Paulson.

was a continual statement. Intellectual honesty demanded that they live as they believed, so feeding their birds the opossums and squirrels killed on the road spoke against waste; their home declared a contempt of materialism; their choice to be outsiders to academia proclaimed their anti-bourgeois values. In a paradoxical way, Hammy's almost obsessive meticulousness freed him to lead the unconventional life so necessary to his well-being. Still, I've always wondered, why he was so silent. You must find out."

Peter Kramer is now a biologist, the head of the International Division for the World Wildlife Federation based in Gland, Switzerland. He honors his father, Gustav, for promoting the year-long exchange. It gave him the experience of a "kind of relation with nature that is rarely available in Europe where we have no 'Wild.'" From the midwestern rusticity of Plainfield to the camp in Arizona where he hiked the Grand Canyon, learned to lasso and shoot bow and arrow, he experienced conservation. "It was something special. I knew then that it was for me."

Plainfield school, however, took him aback. Where were the rules? Here he rode a yellow school bus, and instead of Latin, Greek, and French, math, literature (Heine and Goethe), he had the eighth grade curriculum: typewriting and sports but no foreign language. Peter could beat anyone running. "But I was no good at all those ball games—softball, baseball, football, basketball. No matter; that year was my authentic English class."

There in the amazing great house, he felt Fran's mindfulness, Hammy's quiet authority. Should the boys do anything stupid, a few forceful sentences in that calm, deep voice set sensible boundaries to the unaccustomed freedom. Peter loved shooting; they bought him a hunting license and gave him precise instructions on carrying and caring for a gun, and on behavior in the field. After a few sessions of supervised target practice, he and Alan would grab the shotgun whenever they wished. They hunted, mostly squirrels and rabbits, and often shot pigeons in the barn as practice for the all-important wing shooting.

Peter skinned the squirrels, treated their skins, and sewed a huge pair of mittens for his mother, who came for Christmas. Occasionally he hunted with Hammy and Fran, in his view the "more ardent" of the two. They let him drive the Jackrabbit, the two-seater Ford roadster

Peter Kramer at fourteen, mid-1950s.

with a rumble seat, though not on the highway. He went on excursions with the owl in the back of the family car. They watched for road kills, and on occasion, they simply threw their find back to the bird.

He noted that few neighbors called, but that plenty of "booming friends" were welcomed in the spring. He noticed and learned much within that circle: particularly the interactions of the two of them—Fran asking for approval, deferring to Hammy. He thought it a bit of a comedy—she, exaggerating, and admitting it, he responding in a humorous, self-deprecatory way. "It was an effective way to get to people in a dramatic rather than an objective mode."

They were important models and promoters. Remember, the environmental movement didn't exist in Europe and America, except perhaps in the academic fringe, and in and around naturalists like Leopold. What I learned from Hammy was that there is a way of not taking oneself too seriously without belittling one's message. That is part of any mature person. What was remarkable about Hammy was how he shied away from the limelight. This left him free. There was no trace, no suspicion of self-aggrandizement, which puts him a hundred times ahead of some of the popularizers of today who personalize their accomplishments.

Next, his brother Lorenz came while eighth-grade Elva was in Germany. For him, outdoor life was the thing. No fourteen-year-old could hunt in Germany; here he hunted rats in the corn shed and starlings to feed Murphy the owl. He helped catch mice too, with a flashlight, grasping them with bare hands as others harried them. Elva's rooster sometimes attacked him, so Lorenz armed himself with a water pistol. "Oh, that rooster," noted Elva. "It was a mean, nasty creature. I left for Europe before it matured, and without my tender care it turned into a vicious bird. Ruth Hine still bears scars! I admire my parents for putting up with it while I was gone."

Physicist Lorenz credits that period with strengthening his habits of observation and a giving him a much wider view of people and possibilities. He saw Hammy "become courtly" at some of the more elegant parties with the lively Tympanuchus group; and, when the family spent Christmas with Davis in Roscoe, New York, noticed how genial and relaxed Hammy became. "There was a beautiful camping trip to Texas. We saw the red canyon. . . . Yes, that's it, Palo Duro." Hammy displayed an ironic, spoofing humor, teasing him with tall tales of cowboys. "We laughed a lot." They visited a falconer in Colorado and Pike's Peak, camping every night. He saw falcons and peregrines flown, and a man with a golden eagle. Then he and Alan went on to Vermont, and a wide-ranging tour of New England ending with a visit at Posey's before their return to Plainfield. Lorenz then traveled alone back to New York for his steamer to Germany.

Two years later, came an eleven-year old (Elizabeth Kramer, who preferred the name Eli) and an adolescent (Silvia). Both girls rode the bus to Plainfield school. Elizabeth fell right into Hamerstrom activities. During the fall deer check she climbed up and scooped kidneys from the carcasses strapped on the tops of cars. "I was blood up to the elbows!" Occasionally, she went with Fran on the trapline.

By this time, Fran was deep into hawking. Eli helped put the bal-chatries in order for the use of the Rockford Bunch, the young, rowdy hawking enthusiasts who were beginning to frequent the homestead. According to sharp-eyed Eli, "They were mostly young men in love with Elva." In the spring, Fran sent her out to check hawk nests, and praised her: "Your brother Peter and you are interested in birds. Lorenz didn't know much."

Eli was allowed to train the hawks, a difficult and concentrated effort, requiring enormous patience. Fran taught her the right knots for the jesses and left her in charge when she and Hammy attended a distant meeting. On one "terrible" day, a hawk escaped because of her imperfect fastening. "Fran was very severe, but I think she saw there had been too much responsibility on me."

They were always busy: cutting meat, looking for road kills, doing dishes. Eli did some cooking and in the booming season, she learned to scramble eggs. There were more singular requirements. "Hide that cat carcass!" Fran would say. If the lady who came to get the laundry ever saw a dead cat, she might quit. "So pay attention, or you'll have to do laundry!" When Fran decided the house was dirty, they all (Hammy too) cleaned together in one burst of activity. The children fed Elva's ferret and filed Minerva's talons since they could hurt a human head if she landed there. Should the mice escape from their cages in the kitchen the children had to catch them. Elva named a particular mouse and put it in a special box. It escaped and Elva cried. Eli saw Fran's disapproval

Elizabeth Kramer (Eli) in 1958. She was given such tasks as scooping kidneys from deer carcasses, washing dishes, scrambling eggs for the boomers' breakfasts, and cutting meat to feed the hawks.

of the tears: animals were useful and interesting, but not something to cry over.

Eli worried about the owls perching on the hot stovepipe after Hammy started the stoves in the fall. "Never mind, they'll learn!" said Fran. When Eli tried to shoo them away, Elva slipped off a shoe and threw it at the owl to dislodge him from the dangerous perch—a shoe was always handy and reasonably soft.

The few rules were inviolable. No chewing gum. No slang. No comic books. No eating on the sofa, on and under which Fran stowed her papers. If there weren't enough places at the table, they filled their plates and found a seat near the fire. Children were not allowed in the parents' bedroom.

But in early April 1959, a tragedy interrupted her idyll. Eli's father, Gustav, died while collecting wild pigeons in Italy. Fran and Hammy got a call from John Emlen. He had read a newspaper account of the death of "someone named Kramer." They decided that such a report did not give them authority to distress this child, perhaps mistakenly. Soon after, returning from a short trip, they found Eli on the sofa, looking pale, almost ill. Harry Croy's daughter, probably on the basis of the same report, had written her a condolence letter. And then Neni's telegram came. "Send Elizabeth home right away." They prevailed upon Neni to let her stay.[5] When the time came for Eli to go home, Fran arranged an excursion to Vermont. There she visited with a migration expert who had known her father, then went to New York and sailed home.

Eli had adopted a starling, Ronnie; she wanted to take him home with her. Fran saw no difficulty.

"But how can I feed him?" asked Eli. He needed hourly feeding.

"Why not just take food along?"

"But what will I do on the plane?"

"Keep him in the purse. It's only a few hours. You can feed him in the bathroom."

"And on the boat?"

"Carry him aboard in your purse. In the cabin, you can fly him. If you are careful, you won't need a cage."

She managed; when the fledgling learned to fly in Vermont she rigged starling jesses to confine him. He happily stayed on top of the cupboard when she was not in the cabin. He lived with her in Wilhelmshaven and then in Heidelberg, where he escaped. "Peter was careless," said Eli matter-of-factly.

Silvia, on the other hand, was primarily interested in boys, though reportedly engaged. Fran (whose own books confess her interest in boys in her late teens and her lack of respect for teachers) was horrified to hear that Silvia wrote letters home in German—during class time.

"You must not do that, Silvia. It shows disrespect for your teacher. Promise me you will not do it again." Silvia promised. Then a teacher reported, "Silvia is still writing letters in German in my class." Fran confronted her.

"But they were to my mother!"

Fran quoted a frequent retort of Silvia: "I don't must!" More worrisome was the nearsightedness that made it unsafe to allow the girl behind the wheel. They did let her accompany Bobby Hill, a trusted neighbor boy, in one of their cars. The morning after the outing, Bobby Hill arrived. "Can I see Mr. Hamerstrom?" Fran knew that something serious was amiss. "Silvia drove awfully fast and I hate the way she takes corners," reported Bobby.

The Hamerstroms wired her parents to say that they could no longer be responsible for her, and could her family arrange something else? Adalbert Ebner wrote back: "Your good advices, especially those also of Hammy, will be transformed into previous measures of which Silvia shall be duly informed."[6] He arranged for her to stay for another semester with Leopold's daughter Nina Elder, in Columbia, Missouri. The Elders, with two teenage children of their own, enjoyed her, and Nina still maintains their connection.

Elva thought Silvia to be slyly manipulative. "Besides," she added, "Fran was almost prudish about teenage behavior in those days." Certain of the young male Hamerstrom helpers—with no experience of responsibility for other people's children—have expressed sympathy for Silvia. Fran, who later insisted that children were most successfully exchanged in the very early teens, was sure that they had acted wisely.

The journey to adult independence may trouble even the most respectful mother and father. Alan, the adored "man child" distressed his mother as an adult when he decided on a business career, which has brought him satisfaction and success. Fran talked to me about it. "Alan wants to be a yacht broker!" she said, tragically. "We wanted to save the world, and he is going to be a business man!" After the many such complaints from conventional sixties parents with counterculture offspring that were heard everywhere, this was an ironic twist: the progressive Hamerstroms unhappy with a child's decision to follow a conventional career.

They temporarily disapproved Elva's choice of husband. She met Dale Paulsen at the skating rink as a teenager but did not go out with him because she was not sure how to handle an "older man." When he persisted and courted the quiet, beautiful college junior, they became engaged. To her surprise, and to theirs, too, her parents, like Fran's father before them, assured Elva that it was not too late to change her mind. They timed it better, however, about a week before the wedding.

A chagrined Fran and Hammy later confessed their horror at their first reaction: "It had happened to us, and we didn't like it," Hammy said. "You'd think we'd have had sense enough to avoid it happening to our children." "We love Dale!" chimed in Fran. "I can't imagine why we didn't see him clearly right away."

Closing ranks, they hosted a unique wedding reception on 27 June 1964. They invited everybody: the navy veteran who delivered gas, the garage mechanic who repaired the Volkswagens, neighbors, local friends and supporters, the Madison DNR crowd, and many out-of-town guests. Jim Hale remembered being greeted by a smiling Hammy. "Jim! Would you like to see Elva's wedding present?" And he led him across the back yard to a second outhouse, installed by a colleague in Wautoma, so that there would be one facility for men, one for women. "Hammy enjoyed that mightily." Pink spiderwort and wild geraniums bloomed at the edges of lilac thickets. White-haired and dignified, Posey wore a beautiful blue gown. The Plainfield Homemakers Club provided a fine buffet spread on tables on the sunlit lawn. Owls, tethered on the stumps of the dead elms, looked gravely at speakers when greeted, and sometimes acknowledged their presence by ducking their heads and blinking courteously.

Dale told me of a later time when he and Hammy, driving somewhere together, returned to the subject of their marriage: "He apologized for a long time—half an hour!" Dale shook his head: "He reiterated that they never expected that they would themselves do that to a child. 'It simply shows how inappropriate it is to be judgmental,' he said."

Fran's parents and Hammy's father, outspoken about their hopes, had wished, respectively, for a splendid marriage for her and a brilliant business career for Hammy. Their offspring fell into that very common pattern of self-generated, and often unexamined, expectations for one's children. Each set of offspring did just what it preferred; Fran and Hammy had simply replicated a far-from-uncommon parental reaction to the paths their children take.

Hammy lived his apology to become Dale's frequent companion and genuine friend. Fran was proud of her daughter's career as a wildlife artist and asked her to illustrate several of her books. In Fran's old age, she relied upon her son's good services in emergencies. But Fran, still needing to prove herself and perhaps projecting those needs on women, required a lot from her women relatives. Although she loved to arrange happy experiences for other people's children and to have a creative role in their development, she could not get past her own strivings and conditionings with her daughter, and, sadly, her granddaughters. It was with other young people that she was to transcend her unrecognized needs and create remarkably creative relationships.

# 15

# Of Hawks, Humans, and Freedom

*To band a bird is to hold a ticket in a great lottery.*

—Aldo Leopold

Hammy and Fran sometimes had to choose between their responsibilities to their jobs, which Hammy could not countenance, and the demands of their current enthusiasms and interests. Their semi-official activities in support of prairie chicken–directed organizations—the Society, the Prairie Chicken Foundation, and the Prairie Grouse Technical Council—brought them obligations as well as long-term friendships. But their wishes to play a significant part in such groups as the Raptor Research Foundation Board, the North America Falconry Association, and the Wisconsin Society of Ornithologists could only be satisfied by submitting to constant pressure and frustration.[1] Then there was his editing.

"No one," maintained Fran, "ever realized how much time and effort Hammy put in to improving other people's papers." Ruth Hine testified that he edited departmental material from the very beginning. "He was orderly, disciplined, and persistent," she declared. "Working with him was an intellectual and emotional challenge. Sensitivity was required. I always looked forward to his written or telephoned query: 'Isn't it time for a story conference?'"[2]

Former gabboon Joe Schmutz, now an academic in Canada, watched Hammy spend hours on other people's papers. "His credo, of publishing only when satisfied that the material is of top quality and scrupulously edited, required his prompt, pungent comment and useful suggestions. He never sat on manuscripts for months." Joe's wife Sheila added a knowing restatement of the "publish or perish" formula: "Such attitudes may put those he trained at a disadvantage in the current university climate of publish and publish."

"When you did a paper," said former gabboon Mark Kopeny, "you had to do it right. I wrote half of my Ph.D. thesis up here—on the porch, or in the cool fruit cellar. He was always encouraging and delicate in review. How we waited for—and valued—that nod of assent!"

Hammy even carried papers for review or editing with him on vacations. His letter to an author from their campsite in Manitoba where they were trapping hawks (their regular fall study vacation for several years after 1964) included sincere compliments, an apt diagnosis, and a warning that his comments might be impolite. "Data, based on only five days of observation, is thin." Pointed suggestions followed: "Why not use a little showmanship instead of presenting it deadpan? . . . your readers can't be expected to know that this is red hot." Build from observation to generalization, he advised, and try for clear language: "You'd do your critics (and your final readers) a favor if you could recast the writing into a simpler and more direct style. I realize that you have been indoctrinated (the field marks are unmistakable) and should not be criticized for it, but if you have something worthwhile to say, good English won't hurt it and if you haven't, pseudo-scientific lingo won't help it." A final face-saving disclaimer softened the critique: "Do by all means try this out on somebody else."[3]

The growing nationwide interest in raptors brought a continuing flood of papers to his mailbox.[4] He refereed papers for the *Auk*, the *Journal of Wildlife Management, The Wilson Bulletin*, and numerous proceedings while continuing to write papers and edit Fran's books.

Comparing their bibliographies is useful. Hammy's is straightforward, entirely professional. He and Paul Errington authored a paper in the first issue of the *Journal of Wildlife Management*, and from 1934 on (sometimes as junior author) he published one or two papers a year until 1938. A steady flow of technical papers on various subjects ensued, until the 1960s when his production slowed. His early, classic papers were well known, and he published scientific works as late as 1986.[5] A colleague estimated that Hammy had over eighty publications, plus a number of reviews.[6]

Fran, on the other hand, listed over 150 publications of every kind in her bibliography; titles vary from: "Time to Cook Grouse" and "Dead Trees Are Part of Our World" to "A Morphological Comparison of Two Harrier Populations." She classified forty-one as technical, forty-two as popular, and listed ten books and considerable miscellany—reviews, committee reports, and the like. By her count, seventy-four publications dealt with birds of prey. Strong evidence of collaboration derives from their joint authoring of thirty-five scientific papers, with Hammy as the senior author in the earlier years, and a gradual shift to Fran's name sometimes leading his on the one or two papers or chapters of books they produced each year. "Hammy always edited my work, rigorously," she declared. "And he was demanding. He might say, 'It's choppy.' Or, 'No transitions.' Or, (sternly), 'Did you look up all the references?'" Was she touchy when Hammy edited her work?

I really don't remember. One of us always wrote first, usually Hammy. Then I'd read it. I might say "Darling, you left out the most important thing. Let me write a short beginning paragraph to spell it out for people." Hammy thought of most of the titles.

If either of us ever used undistinguished English, the other would jump at it. We were interested in the product; we didn't disagree about the process. We had different styles, of course. Hammy wrote slowly in intricate patterns. I write fast. You know the "Biography of a Dancing Ground" in *Sharptails into the Shadows*? I wrote that alone. Ruth Hine and Hammy were sitting right at that table. Ruth said we needed a kind of case history, or something to pull it together; I went to the sofa and wrote it.

Ruth Hine agreed. "While Fran wrote that often reprinted section, we tried to think of a title. Hammy suddenly said, 'How about *Sharptails into the Shadows*?' It didn't take us long to see that it was just what we wanted."

Dan Berger has a vivid image of Fran running in, waving an acceptance, and saying, "I've written my last scientific paper!"[7] After 1970, when *An Eagle to the Sky*, one of her best books, appeared, her heart was really in her wish to educate the general public—a large public. She would say: "I want to write a blockbuster!"[8]

After a publisher accepted a book came the drudgery. Hammy became her administrative assistant as well as her editor. He excelled at the detail: the copyrights, the photographs and permissions, and the proofs, but she kept what financial records there were. His stance was explicit: "She supported me through all those difficult years. Now I intend to do the same for her." His help allowed her to indulge her passion for raptors.

Falconry, at first an avocation, was increasingly important to her. She began to keep hawks in 1957 or 1958, as we know from Elizabeth Kramer's memories. Indeed, she was listed by the North American Falconers' Association (NAFA) in 1942. *Birding with a Purpose* details many of her early trapping experiences with hawk owls in Canada in 1962, and of other owls.

Owls fascinated Fran. There was always an owl in the large cage they created by screening in the entire front porch of their house. Jim Hale described her early "trolling" for them, an effort in which he participated once or twice. "She fitted an old rubber overshoe with nooses as a sort of bal-chatri, tied it to the fishing line on a rod and reel, and dragged it down the back roads out the window of their slowly moving Volkswagen. She caught several great horned owls that saw the moving object as prey and attacked it. The neighbors who saw this developed grave doubts about Fran's sanity, or so I was told."

Fran loved trapping; she found the suspense of waiting while the bird footed the nooses or dove into the dho-gaza net almost unbearably exciting.[9] Intense spurts of snowy owl trapping occupied at least three of her winters. By then Hammy had improved the traps she had devised; their modification of Dan Berger's modernized bal-chatri— essentially a bait cage with nooses—could be set wherever owls were sighted: city dumps, breakwaters, suburban neighborhoods, or ice-covered lakes.[10] Banding was no longer confined to migrants or nestlings. Adults could be captured on their breeding grounds. Hammy provided his new Volkswagen rather than his presence on most owl expeditions. It went through soft spots in the ice three times, once with one wheel, once with two, but it was never submerged. Such excitement did not deter them.

The next step was her work with eagles, which went from training to experiments with breeding and artificial insemination. This was new and demanding; it required considerable patience and brought many disappointments. *An Eagle to the Sky* is a retrospective on those

Fran and one of her eagles, in the early 1960s. Neither snow, rain, nor camping kept her from flying her hawks and eagles.

efforts. It makes real the demands of such an avocation, including her emotional investment with her pets, or subjects. Her panic when, with precious eagle eggs in the new incubator they had finally rigged up, she saw the light bulb flicker and go out, shows how involved—and tense—she was. She rushed into the house demanding that someone hurry into town to buy a new twenty-five-watt bulb. Hammy picked up the bulb and looked it over. The filament was intact. He said, "Of course the incubator has a . . . " As he said the word *thermostat,* she realized. The bulb was supposed to go off and on. A long series of efforts to breed or artificially inseminate eagles proved unsuccessful, but Fran learned a great deal.[11]

In *Birding with a Purpose,* Fran revisited another side of her early trapping years in her spirited account of trapping with the rowdy Rockford bunch. This was the group who appeared at the door sometime in the late 1950s and invited her to trap eagles with them. She accepted and then organized them ("I have found that once an order is given, it is best to turn away.")[12] for excursions to Petenwell dam where eagles scavenge the fish mangled by the turbines. Fran found that the boys, some as young as fifteen, were surprisingly aware of rap-tor lore; her memories of her adolescence helped her accept their wild ways, and a serendipitous cohort was formed. The teenagers learned method, rigor, and the value of good manners; she found them both amusing and useful. The book's rollicking tone may hide from the reader the expertise that fills it: methods, tools, and difficulties become clear as her tales instruct the reader, almost unwittingly, how not to turn around on snowshoes or how to trap and manage the live mice, starlings, and pigeons used for bait.[13] Her stories, mostly lighthearted, accomplish two things: they illustrate principles of and tools for suc-cessful trapping, and they demonstrate the depth of commitment and flexibility needed to be a birder with the purpose of learning about the lives of those they trap.

Presently Hammy became deeply involved with raptors too. Hawks provided a kind of busman's holiday from the grouse project. Even in the 1950s he accompanied her when he could to the Cedar Grove hawk observation station near Lake Michigan, where Dan Ber-ger spent fall and spring days.[14] That station is still active. Here volun-teers still watch for, count, and retrieve trapped hawks, which they band and release. They bring tents and supplies; paths and utilities are minimal.

Hammy brought his helpful, thorough presence to all activities, in-cluding joining in the rebuilding that followed a destructive fire that burned the station to the ground in 1958. Indeed the whole family, in-cluding Elva and occasionally Alan and his fiancée, Helen, met in the early 1960s at Cedar Grove several times in the fall migration season. After 1951, the practice at Cedar Grove was to use dho-gazas, an an-cient netting technique developed by Arab and Persian falconers; the couple learned of the practice there.

Soon Hammy was funding all their private raptor activities. He grumbled, not entirely in jest, "It would be easier to give Fran yachts, diamonds, and furs rather than this houseful of gabboons." Although he had no desire to practice falconry, his support was invaluable at key times. Kent Carnie, now manager of the Archives of American Fal-conry, in Boise, Idaho, remembers the significant role the couple played in getting the sport's practice legalized, as it is now, in every state.[15] Fran was the hard-working chairman of the legislative committee of the National Association of Falconers of America from 1963 to 1970. Indeed, says Carnie, she gave up flying the birds for those years in order to concentrate on the legalization effort.

Falconry, little practiced in those years in the United States, was simply not considered when management agencies listed legal hunting methods in their regulations. Then, in the late 1960s, as populations— particularly of peregrines—fell, well-meaning groups tended to make falconers the scapegoats. To counter this simplistic approach, the couple made a spontaneous visit to Carnie, then in the service and a member of her committee, on their way to Mexico. They, and one other committee member, sat and talked under a cottonwood tree at Fort Bliss. They brainstormed, and then wrote a set of model state regulations. Fran, who had prepared an illustrative map, silhouetting legal states in black, served as both prime source and scribe.[16] Hammy brought his lifetime of experience to the discussion and acted as editor. Carnie noted his clear sense of what was attainable. "He was a marvelous help."

Subsequently, draft in hand, they called their old friend Larry Jahn, now secretary of the Wildlife Management Institute in Washington. The Audubon Society intended to present a proposal for a moratorium on any take of large falcons at the Institute's upcoming North American Wildlife Conference symposium. Jahn placed them first on the program, and NAFA's legislative committee's proposal upstaged the proposal to ban. Eventually, after moving through national and international agencies, the work done by Fran and her committee became the framework for America's first federal falconry regulations.[17]

The peak of Fran's fascination with raptors came with the Harrier Project—a twenty-five-year accumulation of data on hundreds of nestings and color-marked individuals. Sparked perhaps by her pleasure in watching the sky dance of harriers on the Leola marsh in graduate school days, her work on this project resulted in one of the very few long-term studies of a population in its home territory as well as her single trade book to feature scientific data and speculation. That book, *Harrier, Hawk of the Marshes: The Hawk That Is Ruled by a Mouse*, is the very footprint of Fran Hamerstrom: her persistence and practicality, her curiosity, her innovative approaches, and her self-confessed foibles.

Fran's prologue in that book announces her discovery that when voles are numerous, harriers abound. Older male harriers "tend to indulge in polygyny," and even young males breed. She identified her purpose, which was to trace her " voyage of discovery." The reader who follows that twenty-seven-year voyage of questions and their discovered answers about harriers learns, as did she, unexpected information. In sum, the varying abundance of voles regulates the number of breeding harriers and determines their mating system. Are harriers monogamous? Hardly. Bigamists and trigamists abound. Do variations in harrier population relate to the four-year vole cycle? Her complex inquiry would eventually address the factors governing harrier mating: nest-site fidelity, mate fidelity, courtship, productivity, nestling development, food items, home ranges, population indices, migration patterns, age structure, agriculture practices, and interaction with prey species. Fran's conjectures and conclusions led her back to her firm opinion that land must be managed for species survival. However, she ended her book with a modest statement: "We are not ready for a recipe. I have presented ideas, and I hope that creative minds will implement them, so that there will never be a spring when the harrier . . . no longer dances."[18]

It all started with her wondering if harriers mate for life. That meant devising successful trapping techniques, and it took two years of experimentation before she began to use the dho-gaza.[19] Dan Berger supplied old specimens of these nets, long used in the Middle East for trapping birds, from Cedar Grove; she turned Ambrose, her pet owl, into a bait owl.

To fulfill her purpose she had to complete a long-term census of resident and migratory harriers, and it became clear that in considering what the harriers ate, she must also make an accurate vole census on her study area. Voles were another cyclic population, this one with a four-year cycle. Hammy took care of arranging leases on the plots selected and helped her arrange to hire Paul Drake. Gabboons helped her conduct "over twenty-thousand small mammal trap nights."[20] In another stroke of good luck, visiting scientist Daniel Q. Thompson was coincidentally creating a vole index by counting the stems cut by voles on sample plots in three states. He volunteered to help her after learning what she was doing. "If you can find a good man to run my transects here, and we can get permanent leases on three 40-acre plots of good vole habitat, I can add the Buena Vista marsh to my study."

The resulting amount of data gave Hammy what he needed to supply the graphs, tables, and photographs that add substance to her book—material that in her preface she characterized as "original data, and dry technical material—not of interest to all."

The ensuing book, beautifully illustrated by Jonathan Wilde, is indeed a hybrid. The text consists of narrative and human-interest stories interspersed with definitions and clear explanations of methods and innovations, which included radio tracking. She used two gabboons as trackers, one with binoculars to flag his companion, should he be off-course in finding the nest. She also devised a quick sampling technique, one that required a mere 120 mousetraps. They based vole population figures on this sample and called it the Hamerstrom Mouse Index.[21] Another practical innovation resulted as she tried to deal with the drastic drop in harrier numbers (caused by DDT from 1960 to 1968) that almost ended her data collection. She wanted a quick biopsy technique and simply decided to sharpen and use her manicure scissors on unanesthetized hawks. The first one she did this on went right back to its usual pursuits—feeding its young, or dive bombing an owl—so she continued the practice.

One of Fran's most disarming traits was her ability to make her many false starts or tensions seem part of an upbeat process. Never ashamed to admit that she did not know enough, she made her temperament, her impatience, and her repeated reliance on Hammy to get her out of a scrape, into mini-dramas. Here is Fran, instructing a gabboon.

"Look," I ordered in exasperation, "you can't just give that rivet little taps. You've got to whang it home."

"I'm afraid I'll break its leg."

"Well, you'd better not!"

"Fran, I wish *you'd* do this."

"If you're ever going to amount to anything, you'll do it." By now I was shouting. "You won't have me around as a nursemaid for the rest of your life."

Frederick gave a little cough. He does not like shouting.

The gabboon, his chest glistening with effort, crouched, hammer in hand, and cast Frederick an imploring look.

Frederick's voice is always quiet. "Just a moment. I have an idea." Frederick rummaged in the tool box, found a pair of vise-grip pliers, adjusted them, and snapped the rivet shut.[22]

In spite of a lack of funds that limited the lab tests she needed, she managed to get enough data to suggest support for the growing suspicion that pesticides were likely factors affecting bird populations. Joseph Hickey's well-known work, *Peregrine Falcon Populations,* advanced this hypothesis.[23] His symposium of 1964 fueled the effort that brought Wisconsin to ban DDT in 1969, a year or more ahead of any action on the national level. Fran delighted in pointing out that eight of eleven of the people Hickey invited to the symposium were falconers.

Meanwhile, she began a project with kestrels. In their first twenty years, she and their workers had found only three nesting kestrel pairs over the entire fifty-thousand-acre study area. So in the 1970s, she put up fifty nest boxes and maintained them for five years. Eight to twelve broods a year fledged from those boxes, hatching about two hundred birds. In contrast, five young a year came from the natural nests. Fran stated her modest conclusion: "The limiting factor was lack of suitable nest sites," in the groundbreaking paper, "Management of Raptors."[24] It pleased her that the concept of managing and preserving these "predators" reached a significant public. "Before," she said tersely, "they just shot 'em."

One thing certain emerged from Fran's work with hawks. It required their continued reliance on gabboons. That their house became a combination academy, salon, field station, and clubhouse—and for some few—almost a chapel, was the outgrowth of the harrier project. It required that gabboons become a permanent part of their household: over the years they had had nearly one hundred such helpers. She officially awarded the title "gabboon" to sixty-seven of them whose names are listed in her dedication to *My Double Life.* They had, she said, "struggled beyond the call of duty on hawk and owl research," but she often laughingly referred to gabboon status as "the lowest form of life." They came from varying backgrounds and several countries: Germany, Canada, Mexico, and Italy among them.

In spite of his unassailable patience, Hammy found some of the high jinks and carelessness of young males somewhat trying and

Two gabboons of the 1990s—David Sikorski and Nora Delia Lopez-Rivera. When asked how he liked working with the Hamerstroms, David replied, "Nothing but good can come of this."

looked forward to a time when there would be no helpers underfoot. The first female gabboon, Deann de la Ronde, was "a refreshing change."[26] Fran, who cared for and directed the motley crew, said simply, "We had six months each year without them, after all."

Being a gabboon was not for everyone. Refreshingly, John Toepfer, who now carries on continuing prairie chicken research in the area says, "I escaped that!" But others attest to deep and long lasting influence,[26] and some parents envied the way their offspring accepted and celebrated the Hamerstrom influence.

Fran chose the gabboons: several were of the Rockford bunch, some of whom became lead gabboons, others matured to become falconers and conservationists. She recruited likely candidates at meetings. She interrogated everyone. "Have you ever had a wild pet? Do you hunt? Trap? Have you made any natural history collections? What do you want to do with the rest of your life?"[27] Endorsements from the right person helped: John Emlen recommended a young man whose father took him out of the university so that he would not be corrupted by Emlen's teaching of evolution. Fran took great satisfaction in accepting that recommendation. She usually built in a trial period, so those who did not like the situation or could not do the work could leave without upset.

Most gabboons stayed a full summer. They trapped and banded birds, put up or maintained kestrel boxes, located nests, made vole counts, or covered the radio tracking when it began to be used. On the rare occasions when fieldwork flagged, Fran set them to maintenance—trimming trees or even cutting grass under Hammy's watchful eye, doing dishes, running errands, cooking, or cementing the steps into the large fruit cellar.

They learned the essentials. "Write up your notes before you go to bed, no matter how tired. Wear conservative clothing in the field; odd attire offends local residents. No smoking in the ballroom. Never rock back on the chairs." Unspoken conventions soon became clear. Hammy's time was precious, his office inviolable. Parsimony ruled: the piles of used paper and cardboard were to be re-used. One paper towel would do to wipe out the frying pan. Baths took place at the pond in an explicit routine. "Carry a full bucket well back on the bank. Soap there, rinse, and only then swim." Gabboons respected their privacy;

they stayed away when she announced, "We're going to the pond," and they kept out of her office on the hot, airless summer nights, when Hammy and Fran put down a mattress on the little north porch that adjoined it.

She claimed that minimal training led to independent action and that the best teaching occurred on occasions when the neophyte became receptive. So fieldwork began almost immediately, although Deann de la Ronde recalls brief but explicit training before being put in the field. In 1972, at least, both Hammy and Fran were at the "round tables" that took place at the end of the day to assess the day's efforts. "No one was looking over our shoulders. If something failed, we were complimented on a good try—unless we had been careless. Even then, a raised eyebrow from Hammy or a Br-r from Fran was usually it, but it was enough."[28] Whatever the task, from trapping pigeons to repairing a snowshoe, gabboons responded to a variety of demands.

Some needed to learn how to handle public relations, some the scientific method. Often we had to teach how to write concise, accurate notes, or provide practice in simple observation. What exactly did they see? Where? For how long? What actions were performed? Sex? Coloration? Anything unusual? If it was a color marked bird, they must record the color and position of the feather. They often forgot simple observations such as wind direction, time, other species seen. Many of them had shocking table manners. Hammy found it painful. We tried to set a good example in that, and in speech—"Yes," rather than "Yah," "lie" instead of "lay" in the imperative—you know, "Lie down!"

Fran was generous with praise. Though she might tease a gabboon for being "the laziest human alive," she told every one—for years—how, after a morning stretched out on the sofa, overhearing complaints about not finding a certain nest, that one favorite, "got up, went out—and do you know, he was back within two hours, with its location!"

Reminiscences provide the flavor of the experience for those temporary family members. For them the taxing, repetitive field routines and the occasional menial tasks were transformed, through Hamerstrom alchemy, into treasured experiences.

Alan Beske grew from a chance visitor into a gabboon, then a collaborator, and finally a treasured friend. On a snowmobile jaunt north with a fraternity brother who stopped by Plainfield to say hello to the

Hamerstroms, he "fell in love with them and the place." When a spring fishing trip fell through, he found himself at Plainfield on a substitute sandhill crane search. There was Hammy, flat on his back in traction. Fran was away on a brief trip. Beske stayed; Fran came home and put him to work. He spent the rest of his spring break with them. Soon he found himself coming to Plainfield from Madison every week. Once he went on a hawk trapping expedition on the marsh. Hammy warned him: "Don't go down Quarry road."

"I heard him," Beske recalled, "but I knew exactly where I wanted to end up. I set off, though I wasn't really familiar with the marsh and ended up going down Quarry Road. I got stuck. I got so stuck that it took two tractors to pull me out. One of them was stuck there for almost a month."

"What did Hammy say?" I inquired.

"He didn't say much, just very deliberately: 'What road did I tell you not to go down?'"

The Hamerstroms treated each gabboon as an individual. Beske later wrote his Master's thesis in Posey's room. He reported, with a smile, that he wasn't allowed downstairs to lunch unless he had written at least two pages. "It was fall, hunting season, and there I was, imprisoned, inside. But I wrote those pages, and then more before I could appear for dinner."

At its best, then, the gabboon experience was a high-class apprenticeship. The immature grew up; the mature lived, breathed, and dreamed of the project upon which they were engaged. Even the lackluster gabboon met wildlife professionals, often from abroad, and heard the experts discuss their work.

Four years after Mark Kopeny met the couple at the 1974 Great Lakes Falconry Association he was a gabboon—and for the last of three summers, the senior gabboon. Kopeny, now a zoology Ph.D., says, "The underlying intellectual rigor of Hammy's work commanded respect; his persistence—a pivotal trait—helped him to contend with change. He was genuinely concerned about passing on information and insights, unendingly patient, kind and delicate in review, tactful, a great listener. I think he liked the mentor role. He certainly was good at it." Kopeny, a serious student of science, was one who received generous interaction from Hammy.[29]

Kopeny further noted how Hammy tempered Fran's stories, and showed deep concern when a bee sting made her good eye swell shut.[3] He saw the respect with which they treated the departmental staff. He admired Hammy's resolute atheism, the couple's gravitation toward the underdog, and their ability to accept confidences. He said, "I can tell them anything!" He implied that being a gabboon saved certain young men from trouble with the law. Curious, I asked Fran. "Tell me about that." "I'm not going to tell you!" she retorted. "That's why people feel they can tell me anything!"

William Scharf became a biology professor as a result of observing the booming twice as an undergraduate. When he was a graduate student at the University of Minnesota a professor called for three volunteers to go to Plainfield one weekend. "I'll go," he said, "and I'll get the others too." He spent the next summer as a gabboon. It started poorly. "After spending the dusk hours of my first day at the Hamerstroms on top of the Kombi [van], alternating between the spotting scope and reading in the Banding Book, I headed back to the house, leaving the Banding Book topside. . . . I was delegated to tell Hammy with my head hung in shame. His words still echo, "Well, you'd better find it!" Not the abusive response I may have deserved, but there was no question as to the seriousness of my transgression. . . . I still remember ironing the pages of that notebook that had gotten wet before it was found." Scharf changed career plans: two years later, he was in Orkney, trapping hen harriers. Like many others, he learned to write from Hammy. He also learned about life, going as their guest to the summer theater in Stevens Point and listening to jazz background music and participating in stimulating conversations at dinner. He noticed that when Fran called Hammy *Gesichtle*—dear little face—it was to appease. "Tinbergen and Lorenz," commented Scharf, "showed that this sort of behavior maintained the pair bond in many organisms. I should have learned that earlier in life."[31]

Joe Schmutz, born and raised in Ulm, Germany, arrived from Canada late one hot night with two friends who knew the Hamerstroms. Fran found them sleeping near the back door in the morning. "Come in! There's coffee." All three worked as gabboons all summer. Joe's curiosity and careful fieldwork pleased them, as did his statement at summer's end. "How I wish I could go into this profession! Of course, it's impossible."

"How so?" asked Hammy. Joe had been in the non-university stream in Germany. Hammy reflected. "If you can improve your English and earn enough money for your freshman year, you can qualify as a special student at the University in Stevens Point." That is what Joe did, working for months in a mine and then living with them the next summer and commuting. Thereafter he found a way to earn board and room in Stevens Point while he got his degree.

In this rustic life, he thought, etiquette would not matter. He soon reconsidered. He saw the stern protocols; noted Hammy's dislike of interruptions. "One time Hammy talked to me, very subtly, about the importance of good manners. If you made them second nature, then at a time of great importance like a job interview, you needn't worry about the impression you were making." Joe said:

They caused me to try to look ever deeper at nature, the process of science, and the interaction between it and the public. Something in the calm, logical, considered way he talked reached me. And you can't escape the honesty. For a required interview about stereotyping in a psychology assignment, the persons cooperating had to adopt a foreign point of view. I asked Hammy to do it.

He gave me an inquisitive look. "What exactly are you asking me to do?" I explained; he took the questionnaire into his office and read it. He came back.

"I'm afraid I can't do it."

"But Hammy, you just *pretend* you are a racist, misogynist, or whatever."

"Sorry, I really can't do it."

"I could have told you he'd never do it," said Fran.[32]

Joe's wife, Sheila now a genetics professor, met him when she was a freshman at the University at Stevens Point. "We had a date in Joe's old Volkswagen, which broke down near Plainfield. He called Hammy. We stayed for dinner—fondue, here on the porch. Fran, thinking I was probably 'girly-girly' handed me half of a dead something and said, 'Go feed the owl.' I passed. Later, we both went up to the University of Alberta, got our masters there, in ecology. When I decided to go on in genetics, Fran felt it was a sort of desertion, but Hammy liked it. He saw the beauty of science in all fields."

Sheila too became a gabboon, and the couple spent their subsequent honeymoon in the marsh house whose windows Joe had glazed.

For years afterward the two regularly came to see their mentors. While Joe and Fran talked hawking, Sheila would visit with Hammy. "He liked women," she observed, "and would give me mending—this button, or that vent on a coat. It must be done right, with matching thread." She might question him. "Does Fran give all the papers?" They were doing the dishes together. Slowly he lifted a plate from the dishpan. "I'm torn," he replied. "The spoken word is important. It can make an impression, be flashy, or showy. Generally, I let Fran do it. I'd rather make sure what I say is just what I mean, so I prefer writing papers."

Dale Gawlik visited the Hamerstrom place from Stevens Point as a lad of twelve and later became a gabboon. Now an avian ecologist for the state of Florida, he directs research on bird communities that can be used in restoration of the Everglades. He believes that Errington and Hammy's views on game management are basic to conservation biology and biodiversity. They valued the intrinsic worth of a species more than its economic importance to hunters or trappers and treasured our heritage and our traditions. Their research of the whole range of interactions between the birds and their environment, "established the factors on which management must be based." Likewise, he said: "Their influence . . . continues to affect every decision I make regarding wildlife, personal goals, work ethics . . . the important things that make me precisely who I am. . . . they instilled in me a feeling about the way I get up in the morning and how I perceive my minuscule role in the world. Hammy felt we were obligated to give something back to the world . . . and treat our short stay on earth as an opportunity rather than a god-given right.[33]

They had always shaped lives. Ray Anderson, close friend and the eventual heir to the booming observations, said, "I am certain that it was from Hammy that I learned it was okay to somewhat explosively indicate displeasure in a given moment. One need not brood about it; get on with the show."

The more one learns about the effort and time required for the avocations and fervor of the activities that now most engaged the Hamerstroms, the more sensible seems their early retirement. They kept thinking of their early years when their passion for the prairie chicken produced remarkable dedication, and when they worked in their own way and at the pace they required. Never stinting, they had put in extraordinarily long hours, ignored weekends, entertained and guided guests, spent unreimbursed "Chief time" on low-level but necessary tasks like lugging heavy traps from field to field. Now, extra hours would not be compensated. And when the state department of administration sent spot checkers—men innocent of the requirements of conservation research—to observe them, the diligent pair felt suspected of shirking.

Perhaps unavoidable, but to them significant and painful disappointments were demoralizing. Mark Kopeny, who had noted and admired the respect they displayed for their department, now noticed tensions. "Questions flew back and forth. I reckon there was just a difference in mentality. The DNR people weren't at that time able to support Hamerstrom priorities."

Hammy, keenly aware of the isolation of the carefully nurtured Buena Vista chicken population, feared that it had become an isolated "mainland island" of chickens unable to find other suitable habitat. He had tried for years to establish colonies elsewhere. He knew that chickens moved between the Buena Vista and a small colony at Mead Wildlife area. A reserve there would limit the isolation; in 1970, the Prairie Chicken Foundation announced that they had the funds to buy land. Hammy, knowing that the University of Wisconsin at Stevens Point could manage the acres, wrote Madison, urging permission to go ahead: "Without it, or some substantial addition to our holdings on Leola, we are licked there if land use continues as it has been these last five to ten years."

The department could not get further involved with chickens. Hammy had to tell the possible purchasers that, with no state budget for management, anyone buying land would have to subsidize management costs too. Hindsight helps: "The emphasis then was on huntable populations; chickens could not be hunted. It might well be different today, when we accept the importance of biodiversity."[34] Then, he was disheartened.

But they would finish the prairie chicken project honorably. Starting it had been demanding and exciting; departure was onerous and slow. The department, pressured by funding agencies, required a publishable final report. Hammy wanted to write a series of journal papers. That would have required years, not months. Checking and systematizing twenty-two years of data and updating field maps was extremely time consuming; eventually, it was only by severely limiting it that he could produce the final technical bulletin.[35] Meanwhile, reminders, deadlines, and the well-meant assigning of others to all duties except "report functions" embarrassed him.

Biometrician Donald Thomson understood the underlying problem. Hammy's data was observational. Such data establishes the context of a species but,

[It] is less fruitful of conclusions than experimental work that can be manipulated and controlled, and used to support inferences. The observational context, however, is what gives inferences meaning.

The difficulties in observational data cut across the whole field. You can't observe everything at the same intensity, and no conscientious person is satisfied with incomplete data. The more you get, the more you generate and test hypotheses. When do you have enough data to come down with firm conclusions? Hammy worked on the broad picture, the whole ecological interaction. He, with Leopold, believed that wildlife management is an art that cannot be reduced to simple formulae. And he did pretty well finish the life history of the chickens. [36]

This imperturbable and patient man hated to remind the agency, which he had served for so long, of the press of his many commitments and of the state of his health. By 1969, according to friends, he had mild emphysema and had twice ruptured a disk. Health was not a Hamerstrom subject of conversation. Should visitors find him flat on the sofa, they would hear only, "Hammy has the punies." He did write to friends: "Our Mexican trip was a great success. Not in the way we had planned it, for it rained every day but two—even in the desert scrub country in which we spent most time—and so I didn't find much sun to walk in. But we banded 49 raptors including such high-quality species as Harris hawk, roadside hawk, and gray hawk. My back is much better now, but I'm still lame and slowed down; I walked a mile and a half yesterday. It seems very strange, after so many years of so much walking in connection with hunting and fieldwork, to set out deliberately to take a walk just to be walking."[37]

Even cutting back to working only 60 percent instead of full time did not help. As he noted wryly, the workload did not lessen. He said, "[It] is very likely that Fran and I will resign at the end of this year."[38] The project was in good hands. Ray Anderson at the College of Natural Resources in Stevens Point was running the booming-ground counts; Jim Keir, successor to Os, continued managing the reserve.

They spent the days of their last leave in September 1971, in the overgrazed National Grasslands in Oklahoma and Colorado. Hammy worked out sensible and modest reforms that would give chickens a chance there: restricting all-terrain vehicles to specified trails; locating fences, campgrounds, and high-use areas well away from natural and scientific areas. "He wrote a wonderful letter, with copies to all sorts of people, to the chief of the Forest Service."[39] It brought little change. The Forest Service responded they were doing most of those things. "We knew they weren't," said Fran.

They had done what was right. They went on to a sharp-tail hunt in the South Dakota badlands and a final trip to Michigan, the site of early effort. Old friend Andy Ammann needed support for his last-ditch effort to buy land to save the few chickens there. Amman arranged a meeting; Hammy spoke of their purchase of seventeen square miles in Wisconsin since 1957 and its success. "Control of key parcels of land" he said, "has permitted maintenance of the unplowed, grassy areas needed . . . [the] species has grown from fewer than two hundred in 1957 to between five hundred and one thousand birds."[40]

They drove back to Plainfield, wrote a retirement letter, mailed it, and left for Mexico. Fran continued the story: "When we came back, we found that Jimmy Hale had not accepted our letter. That let us get a better deal on our state retirement." Then she hesitated. "Don't put that in unless he says you may. I don't want to get him in trouble." But Jim Hale explained, "I just helped him get the most favorable annuity arrangement."

Although Hammy joked about their retirement ("You can't fire me, I quit!"), his graceful response to Kabat's letter of congratulations after the final bulletin came out the next year was sincere:

We were very much touched and pleased by your letter. . . . Now that we are no longer your minions, you do not need to say nice things to keep up morale and so there is an added fillip to your approval.

It did take a long time to bring it off, which didn't please us, either. You were (generally) very patient, and could have killed the whole thing by insisting on an impossible deadline. Again our thanks for letting us see it through. . . .

We look back with pleasure to the days when we were your minions— quite a few years. And I might add, with a great deal of respect for the way you stood behind your men when the chips were down.[41]

The 1973 DNR Bureau of Research award for "a superb demonstration of the translation of research findings to active management," speaks to the regard in which the couple was held.[42]

But no longer would they be circumscribed by the demands met so assiduously for so long. Now they could trap as many raptors as they wished. Fran could write all she wanted. They could redirect their energies. This was made clear on a fine June day well after their retirement date when I went up there to dig a clump of pink spiderwort that she had offered. She left her typewriter and walked out to show me where the plant was. Suddenly she turned and said casually, "We've left the DNR." That was that, but when I asked what they intended to do in retirement she said, sharply and firmly: "We aren't retired! We're just concentrating on our own work."

Now they belonged to a wider world.

# 16

# Free at Last

*Man is made by his belief. As he believes, so is he.*

—Bhavagad Gita

Their routine appeared to be about the same. The back door was always open; Fran would be sitting on the bench by the long black trestle table in her office, typing. He would be at his desk behind piles of folders and papers. They still lived with owls, went to meetings, traveled in the fall, and disappeared in the winter. The themes of their life—public service, pleasures, friendships, instructive travel, and above all, personal fulfillment—remained constant.

Now, however, the frequent exclamation was, "So much to do, so little time!" Hammy, the dean of prairie chicken research still "helped guide many a beginner . . . [to unravel] the mysteries and . . . wonders [of] the grouse clan."[1] He answered every request for information. One, to the high school student author of "A Study of the Conservation Efforts in Connection with the Prairie Chicken in Wisconsin," shows his delicate approach. After generous praise he suggested a few tricks of the trade in citation, explaining, "I have been an editor for twenty years; I can't read a paper without a pencil in my hand. Hope you get an A+!"[2]

He kept on the watch for land for the foundation or society to buy. In 1977 he made offers of $900, $1,000, and $1,200 an acre to a landowner in the town of Carson and was refused. "Illinois prices! Fantastic!" he exclaimed. The next year, he found land available but then, "no $, after a long time with too much in the treasury. Crazy world."[3] He annotated the annual DNR reports, attended meetings where changes in management were discussed. He heard about the

A typical Hamerstrom campsite, this one in Nebraska in 1967. The van has a pigeon cage and gear on the top; the eagle is Chrys, on his portable perch.

spring booming from Ray Anderson. In the late eighties he told Alan Beske of his disappointment in the state of the marsh—too much grazing. He knew that prairie chickens were far from secure: in 1985, in spite of Ammann's efforts, they disappeared in Michigan. He shook his head. "We lost that one."

He maintained close contact with colleagues. He and his former graduate student Ron Westemeier, then with the Illinois Natural History Survey, compared notes from 1968 to 1988. "Sounds as though you have a world-beater of a paper in the making," wrote Hammy. Ron's reports of pheasant predation in Illinois validated Hammy's Buena Vista experience; late-started pheasant control in Illinois and private stocking—"an old story," Ron lamented—made the prairie chicken situation in that state "likely beyond recovery."[4] When budget cuts threatened Ron's funds, Hammy was the one to send Ron's employer a strong message: "From the research point of view, no one can come even close to the data that Ron has on chicken nesting—and these are data from the original range, the very heart of the original range. His data on what an acre of Illinois prairie can produce in the way of chickens are a yardstick of enormous importance. Who else can even guess at it? . . . And his data on what constitutes the size of a workable, practicable land unit on which a viable population can be maintained are also unique. We've all been guessing about that for years . . . you've got a winner: treat him as such."[5]

Both Hamerstroms became faculty and research associates for the University of Wisconsin–Stevens Point in 1972. (In 1982 both titles changed to adjunct professor.) Fran thought it exciting, "Call me Professor!" she exclaimed. He was silent for a moment and then countered. "I believe," he said in his deliberate way, "that 'Professor' was the name given to pianists in houses of ill repute." Hammy, who found intrinsic satisfaction in research and his relationships, was not one to buy into academic status.

Moreover, since he knew the distinction between truth and literal exactness, he had no fear that her popular writing would harm their reputations. He saw how happy acceptance of her books made her and the zest with which she encouraged sales. He knew that she must live each book as she wrote it and jested: "I tell her, 'It's nice having known you,' as she disappears into her study." Indeed, in those years, she had a secret office, upstairs in the barn, furnished with a desk, chair, cot, and electricity for her electric typewriter. Even Hammy was not supposed to disturb her there. If absolutely necessary, he went to the bait room underneath it and called to her through the grain chute. Although her writing was her own, and, "his editing more for polish than for substance, it was a process that still played on each other's strengths and weaknesses. His stability, scientific caution, and courtliness balanced her spirit, wildness and ingenuity," observed his son-in-law.

After fulfilling the scientist's task of writing papers for specialists, Fran had a new mission, one that Konrad Lorenz had practiced, to "reach out to the larger audience and change the ordinary person's view of the world. . . . [It is] the scientist's duty to tell the public in a generally intelligible way, about what he is doing."[6]

All of her books, save one, were aimed directly at the public: a book about her eagles, children's books, hunting tales, trapping experience, their exploits and escapades with prairie chickens, raptors, and gabboons, a wild foods cookbook that still sells, and the autobiography that informs this book.[7] *An Eagle to the Sky* was reprinted several times; Konrad Lorenz introduced the German edition. *Walk When the Moon Is Full* is a children's classic; the exceptional black and white illustrations illuminate both setting and theme. This book gave her a faithful regional following, and, four years later, an enlarged international readership, when a Japanese version appeared.

For the maximum exposure, seven of the brief chapters of the 1980 *Strictly for the Chickens* appeared in magazines, and *Natural History* magazine published some of the early chapters of *An Eagle to the Sky*. Fran's stories of the odd, funny, or startling episodes that had marked their work had entertained their guests and visitors; now her printed collection of those tales led to the August Derleth prize for *Strictly for the Chickens*. Not all reviews were favorable. The director of the library at Lawrence University in Appleton found it "frivolous," the title and cynical, its aloof tone annoying; he felt that she exaggerated the rigors of central Wisconsin and resented her dominating the work. He thought the book contained, "no significant systematic environmental statement." In spite of all that he thought that public libraries should buy it, indicating a possible appeal for the undemanding public.[8]

He missed her purpose—to get the public's attention in whatever way she could, and to move them from interest to sympathetic, if not educated, participation. Fran was not trying to write the mature assessment of the life work of a brilliant ornithologist or a weighty declaration on the environment. She was a realist. "You'll find out what people will read," she warned as she spoke to writers' groups, which she often did.

Those colleagues of Hammy's who hoped he would write the definitive prairie chicken book assumed that he would have "produced," if only Fran had not "dominated him." That charge, easy to make, is hard to substantiate. She did monopolize conversations, especially during the time that she was delightedly publishing books. Colleagues may simply not have recognized the change in Hammy's priorities. He had always known his mind. Joe Hickey asked him once, "When are you going to get that prairie chicken material out, Fred?"

"Never!" he replied.

He quietly made his decisions and went on contributing to his marriage and his beliefs in his own way. He felt no need to justify his personal choices, but as a private citizen, he now felt free to express his forceful, and even controversial, opinions.

Human population must be cut back. "I suspect that man has no more than 10 years at the rate we're going but I'm damned sure we don't have 50. If we don't smarten up pretty soon and make peace with our environment we're done."[9] In 1974 he said: "When there were relatively few people and relatively a lot of wildlife . . . an individual . . . could go out, whack down a hole in the forest and . . . live off the country. He wasn't even making a nick on it. But now, with so many people . . . what one individual does can make a whale of a difference . . . you can see that there isn't much left of . . . things that are of great importance—one of them being space. . . . natural laws . . . apply to us as much as they do to prairie chickens, trout, and deer. . . . if we destroy our environment, we destroy ourselves."[10]

His pronouncements were cogent. "Anyone who does not agree that the world is over-populated is either a fool or an economist." He backed up his words with donations to the causes in which he believed. After his death, Fran remarked, during a gigantic sorting of mail: "Oh, Hammy gave to all those appeals. I throw them out."

The Hamerstrom code mandated helping people. Unlike their way of life and housekeeping irregularities, their good deeds remain largely unknown. There were eloquent letters of recommendation for gabboons and colleagues. Although she never provided details, she might refer to a friend who had been in some kind of tangle, with a tantalizing phrase: "We saved his job!" They were constantly involved in detailed efforts to organize useful visits for European naturalists.[11] Fran often arranged stays with American families for the refugee and foreign children she deemed needy. She invited to lunch, every Monday for nearly a year, a young neighbor whose wife had recently had a severe psychotic break. It was a much-needed respite from anxiety, child care, and labor in the fields. When DNR employee Larry Crawford told them of job difficulties that tempted him to quit and "go for State Trooper," Hammy said, "Don't decide just yet, Larry, let me see what I can do." He straightened the problem out; a difficult supervisor was transferred. Fran found many occasions to act, from providing small scholarships to local high school students for banding bluebirds and

kestrels to giving an antique round table to a couple who had been particularly helpful to their work. Naturally warmhearted, she responded
helpfully to difficulties. Only recently I heard how the wife of a WSO
member who had just lost an eye was enormously cheered by Fran's
matter-of-fact talk with her about life with a glass eye.

Underdogs could count on the Hamerstroms. When Fran heard of
new neighbors at the old Christensen place, she stopped by. She saw
that the Sharkos were refugees. Uncertain of what language to use, she
tried German. "Nick backed up against the refrigerator, trembling in
disbelief; Alley put her head on the kitchen table and wept. Finally,
here was someone who could talk to them. They said that their
sponsor was a retired minister who made no provision for their support. They had a house, food enough for a few days, but no car and
no money. Nick, with shrapnel in his chest, had fainted when he tried
farm work in the hot sun. They were desperately alone. Indignant, she
went home to Hammy. 'Of course, we must help them," he said.
"Damn that sponsor!"[12]

Alley Sharko, an architect's daughter, was soon helping with their
housework. Hammy listened patiently as she recited *A Tale of Two
Cities* in German while he was trying to work at his desk. Eventually
the Sharkos moved to Neenah where Nick worked at the paper mill
and, as a paper company executive at a Tympanuchus Society gala reported, Alley became the "educated scrubwoman" of executive offices.

Some years later, the Sharkos' son, in medical school in Madison,
could not get an important assignment from the library reserve section
because he was outside the group that passed the single copy around.
"Give me the name, the publisher and the date of the book," Fran demanded. She drove to Stevens Point, got the book from the university
library there, and took it to him—a two hundred-mile trip. "Mail it
back to Stevens Point when you're finished. You have two weeks."
Alan observed dryly, "It wasn't always easy to befriend the Sharkos,
but once my parents had accepted a responsibility, they carried it
through."

Their commitments were long term. Injustice must be opposed. In
the case of Rockford bunch alumnus, Jack Oar, that meant a long, expensive fight. In 1974 only thirty-nine nesting pairs of peregrines remained in the lower forty-eight states. In response, the Cornell-based

Peregrine Fund began a program to increase populations by raising captive birds.

The Fish and Wildlife Service believed that there was a damaging multimillion-dollar international trade in endangered falcons and mounted a sting, called Operation Falcon in response. They suspected Oar, an avid falconer. In the summer of 1984 they arrested him after the eight-hour search they called a "routine inspection." Federal agent Richard Stott and four state game wardens ignored his request for a search warrant, confiscated his peregrine and his gyrfalcon, and set out to prosecute him. Agency employees confiscated hundreds of birds of other falconers, including those connected with the Peregrine Fund effort, then based at Cornell.

Oar later learned, through the Freedom of Information Act, that his peregrine had died in Culver, Kansas on 8 September 1974 from "choking while eating." She had been fed one or two mice a day, neither as much nor as appropriate a diet as Oar had fed her, and weighed ten ounces less than her normal weight.

The very concept of a sting infuriated Hammy, as did the eight-hour search and the treatment of the peregrine. He mounted a letter campaign for additional funds and a search for appropriate counsel. Eventually he and Fran contributed $1,000 for Jack's successful defense, which came only after thirty months of procedural irregularities, delays, and pressure to confess or to plea bargain. Oar, who refused to plea-bargain at his trial, received a verdict of not guilty from the magistrate. He mailed a copy of the decision on 24 September 1985 to Hammy, who exclaimed, "At last! An honest judge!" The decision declared that Oar should have been given a chance to show his right to the bird and to get bands, since up to 10 percent of bands do come off.

Hammy urged Oar to sue for the damages to the birds and to his reputation. His name, on an offenders list, had been widely circulated, even to the Chicago Academy of Science. Oar did so; the complaint has never been heard. Fran was able to deliver an impressive, reasoned rebuttal to a government film defending their "sting," at the 1987 Inland Bird Banding Association meeting in Rapid City, South Dakota. Protests by others similarly treated did finally result in a Justice Department conclusion that Operation Falcon "found no one (other than a federal agent) who took a raptor from the American wild and sold it."[13]

The help they gave others was one way the Hamerstroms demonstrated their self-sufficiency. It took a tornado, literally, for them to ask for help. A hop-skip-and-jump whirlwind hit their home in July 1985; they called us. The windmill was down; a tree slanted across the driveway; both brick chimneys were gone; the woods were a chaos of fallen trees, jagged branches, and smashed underbrush. Hammy, with flushed cheeks and hair disarrayed, sat by the blocked driveway in the shade of a sapling. Suddenly he looked old. Fran, slightly rattled, maintained, "I saved my papers; the first thing was the manuscript! Then I checked on my man!" Pete Frost chugged right out with his end loader and cleared the driveway. "They never forgot that. They thank me almost every time I see them." They raised the windmill; Mark Kopeny, a capable and trusted gabboon, relaid the chimneys, and in a few years the woods looked as green and wild as ever.

The Hamerstrom style did not change. They were a unit unto themselves, valuing their privacy and living, with a considerable sense of proportion, in the present. Fran felt no need to advertise their status. Tributes and marks of distinction might appear in the media but were not spoken of. The local newspapers reported their receiving—with Senator Philip Hart and Jacques Costeau—the two-million-member National Wildlife Federation's Distinguished Service Award for Wildlife Conservation in March of 1971. The 1987 Leopold Centennial, where they and fifteen other Leopold graduate students spoke of their mentor, was never mentioned. My question about the long list of awards, prizes, honorary memberships, and certificates they won, individually or jointly, drew a measured response. She thought a moment and then said, "We were very fortunate in their timing. Often, just at a time of great criticism or doubts about the future, someone would come through with an award. I think it helped."

A glossy shot of them came to light in 1993. He looked like a statesman in a tuxedo; she was resplendent in an elaborate gown. What was that occasion?

"Oh, that was on the Queen Mary, on their Big Band crossing. We thought it would give us a chance to dance every night. They invited us to join the captain's table."

Fran, as was her custom, emphasized the professional importance of their trips. In previous years they had explored the range of the greater and Attwater's prairie chickens, and much of the range of sharp-tails and other grouse in the United States and Canada. They had previously managed three trips to Europe (Austria, Finland, Lapland, Norway, and Sweden), to study conservation, grouse, red and roe deer, and hunting ethics. They had spoken at six of the seven Ornithological Congresses they were able to attend. Now they expanded their travel to the Yucatán, to Australia, India, Iran, Siberia, and Sri Lanka, and later to Indonesia, Holland, Spain, and England.

Hammy viewed this wider scene with his customary curious, judicious eye. In Moscow he observed that the capital of "the evil empire" was a flower-rich city with spacious parks, wide streets, and a live-animal market. He noted the heavy foot traffic, sampled the plain, good fare in local restaurants, observed the well-dressed women in heels, and the universal courtesy and friendliness of the inhabitants—not at all what our media had led him to expect:

Southern Siberia is taiga or boreal forest and is much like Northern Wisconsin and Canada. . . . the important thing, the big thing was the similarity. The differences were minor. I daresay it would not be stretching it to say the same could be said for people.

We did not engage in any political discussion and no one attempted to open the subject with us. In Zagorsk we visited an old monastery that dates back to the early 1300s and there a monk tried to sell us on the Russian Orthodox religion. That was literally the only propaganda to which we were subjected.[14]

For fifteen years their travel pattern included driving to their winter haven at Welder Wildlife Refuge, which lies within a much larger ranch and active oil field. They sojourned, wrote, and trapped at the 7,800-acre reserve near Sinton, Texas. The original ranch house, set about a mile back from the fenced entry, is the center of a complex of attractive buildings with dormitories and a kitchen for residents. A foundation has offered support and research opportunities to biologists and students at Welder since 1954. They wrote undisturbed there.

"We have no phone," Hammy told us. "I find that ideal."

Colleagues visited them there, often from some distance. Novy Silva, at Texas A & M, drove "all that way!" said Fran, just to see Hammy.[15] Old friend Art Hawkins went hawking with them once there. "It was hairy! He drove—you know his driving—on a crowded

Texas highway. She perched next to the open side door to toss out the baited bal-chatri. When she did, he went on, slowly, and if a hawk struck, he'd make a quick U-turn, wait in case others might foot it, and then they'd move in quickly to scoop them up and band them. It was hardly what you'd call a relaxing ride."[16]

Hammy was prepared, with slides, for the occasional lecture. On one such occasion in McAllen, Gert and Ray Goult, old friends from the Plainfield Grange, were in the audience. As the image of their unpainted Plainfield home appeared on the screen Fran sighed, "Oh, isn't it beautiful!" Gert laughed indulgently. "She really thinks it is beautiful!"

After Welder they would camp their way to Mexico, to an estuary sixty-five miles north of Kino Bay, where the complete solitude they craved was broken only when they chose to visit the Seri Indian encampment nearby, or Becky and Ed Moser. These American linguists had learned the Seri language and transcribed it into written symbols in order to print a Seri New Testament.

They found the Sonoran gulf-coast desert to be eternally picturesque and interesting. The elephant trees ("*Bursera microphylla*," murmured Hammy) with their red brown trunks and papery bark, rubbery branches, and small compound leaves, give small shade. Cardon cacti, relatives of the saguaro, rise from the almost naked sand. Black vultures roost on the cardons, coating the gray green spires with white. The warm gulf water supports abundant sea life; ospreys flourish. They savored the clear dry air, the black nights and piercing stars above them. "We wake each morning to the cooing of doves. Facing west, we can just make out Baja California on a clear day. Occasionally, we spot a whale. Turning east, there is nothing except sheer desert."[17] Morning and evening they checked nests, walking to some, driving to many more. Hammy made meticulous maps of the nest sites and took pictures of every nest.

They found all they needed there, even medical attention. In March 1972, Dale, Elva, and their twin daughters joined them in Reno and drove in caravan with them to Dove camp. Hammy became gravely ill. He was in misery, with severe renal pain and high fever. Fran rushed him to Hermosillo. There they found Dr. Porfirio Carlos Estrada Aras, who diagnosed Hammy with pneumonia and a serious case of kidney stones. Dr. Aras told the Hamerstroms that they could either go to

Hammy inspecting an osprey nest on a Cardon cactus on the Sonoran coast in 1974. The mirror fastened to the top of the long pole allowed him to inspect the nest's contents.

Tucson for treatment or stay in Hermosilla. They opted to stay where they were. Fran remembered, "The little nurses spoke almost no English and called bearded Hammy 'the heepie,' but they let me sleep in his room." And when Dale and Elva bumped their way to Hermosillo a few days later, they found him improved enough so that they felt free to start for home. When he spoke of his illness in Mexico with Sheila Schmutz, "he recorded the pain of that kidney stone attack and the high fever. He liked the doctor, but not the lack of system there. It was a vivid experience—realizing that he might die. Close partner that he was, he didn't want to leave Fran."

He was in the hospital for ten days after which they chose returning to camp over a hotel for recuperation. "He is most likely to recover outdoors," said Fran decidedly. He did, at Dove Camp. Meanwhile, Jim Hale had a call from a lady who spoke little English. "She and I with no Spanish, worked out the insurance claim which was paid in full through the DNR's retirement benefit program." Yearly, after that, they visited Dr. Aras. "When you save a person's life," said Hammy gravely, "you have a certain responsibility toward him." Their next owl bore his name, Porfirio.[18]

Their camp was near to Seri Indian villages. The Seri, moved to the mainland from Tiberon Island by the Mexican government when the isolated life became too difficult, were still hunters and gatherers in the 1970s. Today, with ironwood carvings they can sell, they have moved into the money economy.

The Hamerstroms liked the observant and responsive Seri and their earth-easy ways. Fran relished the odd humming sound they made as she took the mice, caged and kept for bait in trapping, out of the van. The Seri seemed to accept them, and inspected the pair closely.

"How long," they asked Becky Moser, "have they been married?"

"A very long time."

"Ah!" They nodded approvingly. The Seri value fidelity and did not fail to note the various "wives" brought along by American traders on buying trips.

"Are they very poor?" they asked.

"I don't think so."

"Perhaps they are stingy?"

"They buy very little." Becky Moser explained: "I think they care mostly about birds." The Indians called the couple: "They who capture the doves"; Fran was "the old woman who looks for birds."

The Hamerstroms sometimes ate breakfast with the Mosers. Humanist Hammy ("guileless, a gem") courteously accepted their custom of reading aloud from the Bible, but made his own stance clear with a jest about being heathen. Lexicographer Becky offered an etymological insight that the word *heathen* originated from heath, or uncultivated land. Hammy smiled: "That fits us too. We work hard to keep our acres wild," and acknowledged the logic of her repetition of Pascal's argument for God. Impetuous Fran got up abruptly and headed for the living room. "Becky, that's preposterous." Hammy rebuked her mildly, "Fran, Becky's logic is completely correct."

There they met Ike Russell, a Lincolnesque rancher and his wife Jean who flew their plane into Desemboque on one of the Hamerstrom's visits. Presently the Russell home in Tucson became a regular stopover for the Hamerstroms. Jean Russell relived those times. "They would call. 'We're at the corner of Speedway and I-10!' Then they'd arrive, wash a month of dirty clothes, and tend their birds. The first time they came, Fran baked her famous pies. They insisted, one time just after they drove in, to drive up, nearly to Phoenix to get feed for us at a very busy time. What a lovable man he was. So helpful. When I complained that it took three people to fasten the top of my Carmen Ghia convertible, Hammy sauntered over. 'That won't do,' he declared, and pulled out a phone number. No local dealer had been of the least use, but when I called that number, I got a new top, pronto."

They made new friends at the Russells and celebrated reunions with old ones, like the Emlens, who by coincidence had previously met Ike and Jean in Africa. One time they stayed in Russell's condo for a brief visit, and Fran relished the tone of the place. "Guess who our neighbor was! Burl Ives! And next door was a well-known expert in aquaculture!" They celebrated such friends and such acquaintances.

As well they might. They had remarkable friends. Phileo and Edith Nash were close. After a notable career in Washington, Phileo, a Ph.D. in anthropology and an aide to presidents, became the U.S. Commissioner of Indian Affairs in the 1960s. In 1977, he retired to his Wisconsin cranberry bog.[19] Edith invited the by now well known "prairie chicken people" to dinner. She was, like Fran, a gifted cook, and on her first visit to Plainfield, recorded: "We went to a party at the Hamer-stroms the other night and thought we were back in Rappahannock County. . . . They had four pies on the floor when we came in and three live musicians and a silent poet who played dishwashing music for the dishwashers after supper. . . . The first pie was an hors d'oeuvre—wild mushrooms, in a delectable, outstanding crust. . . . They don't shoot bears but a hunter gave them bear grease . . . very delicate in flavor."[20]

In 1987 Hammy made weekly visits to the doctor in Wisconsin Rapids because he had contracted Lyme disease. That period coincided with Phileo's final illness; they always stopped to see him on their way home. Eavesdropping on the talk of these two men could have provided an extraordinary commentary on ecology, prejudice, the utility of government, and the Great Depression. Or, since Phileo owned a cranberry bog, they might simply have discussed the weather or the state of the cranberry market.

Other friends were Arnold Zimmermann and his wife, Elizabeth. They had fled Hitler's repression in a version of the Trapp family trek. He became brewmaster at the Schlitz Brewery in Milwaukee and a member of the Tympanuchus society. They went booming: "We rose to a pan like this," he held his arms wide, "of scrambled eggs. We saw two copulations and recorded the band numbers—not easy to do. Fred checked our data and said, 'Oh! old number thirty-six!' He knew that bird."

Elizabeth wrote classic books on knitting, a bond with expert knit-ter Fran. Eventually the couple remodeled a schoolhouse near Babcock as their retirement home. Arnold remembers a Hammy that not every-one knew.

You could touch almost any subject and he would respond. He was a partic-ularly good geographer; he knew a good wine. He told us about his illness in Mexico when was so sick he didn't give a damn about anything. He was very gentle, and would bring her around to his point of view quietly. But I faulted him for not stopping her from smoking. I believe he had emphysema. I spoke to him about it, because smoke was bad for him. He shook his head. "She is very thick-headed, you know."

We were both liberal in politics; we shared a love of music. He and I were really on the same tuning fork. I told him about seeing Paul Whiteman when Whiteman's huge orchestra came to a variety theater in Munich with Gershwin. When I came to the United States I worked in Trenton, New Jersey, where Whiteman was born. I joined the New Hope sportsmen's club; we shot skeet. Paul was there regularly. I was treasurer, and he would always give me a blank check. He had a Weatherbee shotgun.

We saw this tall flower in our meadow that had some small white beans as it finished flowering. We thought we could eat it. Betty cooked it up. I had hardly swallowed my first bite when I was struck, immobilized. I couldn't move; I was afraid I couldn't breathe. But it passed. I mentioned it to Hammy—of course he knew it. "Why, you ate wild indigo!" he said. "It is poisonous!'

They led the life they preferred. They hated petty politics. He would not have been happy in a bureaucracy, or even in the university.

Fran sought out musicians, and Hammy made sure they could find them again. His little black book held a record of their musical contacts so trips would include good folk and dance music. Our phone might ring with an invitation. "Helen! We're having a concert tonight. Do come. Paul Bentzen (a local musician) will be playing the spoons." Or, "Dale is going to prepare smoked chicken tonight. Can you come for supper?" They threw a party for the Rockford bunch before Rodd Friday went to Vietnam. Some parties were memorable; at least one went on all night. Practical Fran and Hammy slipped out for a good sleep in a motel as the hour grew late.

They were frequent guests at our house. How they enlivened a party! He, white bearded with full upturned mustaches, was a presence, attentively responding to the interests of guests. She, in a boldly printed caftan, created theater. Once she entered with a young red-tailed hawk on her wrist. To the mixed fascination and consternation of a crowd of university types, Fran asked for a cutting board, pulled out a large functional knife, chopped a scrawny chick carcass into quarters, and fed that hawk before our startled eyes.

No celebration could match the one in Plainfield of June 1981. They dubbed it their one-hundredth anniversary. For her, using romantic if not arithmetical addition, the two ceremonies doubled the anniversary years.

Colleagues helped with the arrangements. Alan and Helen, Elva and Dale came early from opposite coasts; friends put new screening on the porch. Putnam and Dottie were honored guests. "Fran was in her element; she invited everybody!" said her daughter. "I didn't know half of them!" College friend Thad Smith mingled with more than 150 people from Alaska, Arizona, the East and West Coasts, even Germany.[21] A mileage chart showed a total of over one hundred thousand travel miles. Afterward a jubilant Fran wrote Putnam about the "family rally," as she called it: "Your photos are here and they are priceless. [In] #21 you ask what holds my dress up . . . and I reply, I don't know. . . . Plainfield, and the neighbors have been a-buzz. We stick to our story: it was a SURPRISE: we expected a few friends but not over two hundred."

Actually it was very moving to have so many there with whose lives we had been closely associated. Now let me put your mind at rest: the goat is now FAT & giving lots of milk."[22]

Fran, with memories of Mexican *cabrito* fresh in her mind, had planned a barbecue with Dale as butcher and cook. He saw at once that the goat, donated by the Hamerstroms' orthopedic surgeon, was too ancient to cook. Susi Nehls, Joe Hickey's daughter rescued it. "I just took it home with me in the back seat of my Volkswagen bug. As we left, it was calmly looking out the window."

Of course their lives were not free from difficulties, but they did not dwell on them. After their close friend, painter Jonathan Wilde, could no longer spend winters there as he had since 1962, they found other house sitters, with increasing difficulty. One bad year, a local couple agreed to live in the house. The man took his chainsaw across the road and cut the huge live white pines, borrowing Larry Crawford's pickup to haul the wood. "It was green wood," puzzled Larry. "I never could understand it, and they were awfully upset." Their dismay and indignation remained for some time.

But generally, their life was free of strife. True, Fran had sensitivities. Talking to her about smoking was unwise. She simply couldn't give up the long-term habit. Artist Deann de la Ronde saw Hammy whirl and leave the room if he saw her with a cigarette in her hand. He rarely had to do this, for she would push the offending object into a companion's hand should she hear him coming. Nor could she admit that it was bad, either for her or for others. Hammy, impatient with her refusal to concede that the practice might have something to do with the state of her lungs, once murmured to Elva, "They should take away her biologist's license!"

A young woman who once inquired of Fran, "Aren't you ever angry at him?" got a quick response.

"Of course I am! I threw a blue and white bowl at his head once when he wasn't paying attention to what I was saying." She grinned roguishly. "I have good aim. I made sure it just missed!"

She chafed at his caution. All December, looking ahead to the long drive south, he listened to the long-range forecast on the radio. She thought they should simply leave when they wished. "People," she declared "who let the weather control their choices are prisoners of the weather." Hammy had dealt with too much bad weather to want more. An undated note to Ruth Hine from Sinton makes clear why. "Hectic getaway, starting Christmas afternoon, returned after twenty-five miles. Had to be towed to start motor, which ran very badly— probably ice in carburetor. Towed again next am, finally off by noon, running better but no heat (–7 degrees Fahrenheit). Arrived here in rain, followed shortly by small scale ice storm, more rain."

Understanding and mutual permissiveness muted the few differences between them: should they differ they settled the problem politely, or privately. They attested in every possible way to their love for each other. Hammy was mystified by a visiting husband who received his sobbing wife's phone call about a cat's death with a brief dismissal: "Well, honey, you'll just have to get over it!" Hammy said, "If Fran should call me in tears, I would drop everything and go to her at once."

That statement explains all. Their deep attachment had accompanied them through drudgery and adventure, success and criticism, poverty and modest self-indulgence, pleasure and loss. It would accompany them through the final journey.

# 17

# The Making of a Legend

*Sing the great lovers: the fame of all they can feel is far from immortal enough. Those whom you almost envied . . . Begin ever anew their never attainable praise.*

—Rainer Maria Rilke

Through all those years, the Hamerstroms were doing more than saving the prairie chicken and studying raptors. They were creating a legend, one based on solid scientific achievements, their work with prairie chickens and birds of prey, his classic early papers, and the public awareness built by their hospitality and her books. His generosity in advising, editing, and organizing, and her flair for publicity certainly helped. The interaction of these two personalities in long years of collaboration in a loving partnership was key, for each of them accomplished much more than either one of them could have alone. They were a genuine team.

People remember them as such. A resident at Welder Wildlife Refuge, where they wintered for many years, reflected: "It was something to see. There was no competition. They were each other's best friends, always thoughtful and polite to each other. When you think of sixty-one years of working together—it really was inspiring."[1] Neighbor Larry Crawford, a worker for the DNR on the grouse project for eleven years and a friend for many more, never heard disagreement or bickering. Fran wrote of "a cast of falcons, each anticipating or responding to the other's moves,"[2] and the phrase precisely describes the couple's interactions. Words such as decision-maker, supporter, or dominant—words with significant overtones—are a less accurate reflection of them.

They learned from each other and then used that knowledge to get the most effect from their work. Consider their early repair of the foundations at the Walter Ware place: after cleaning, washing, and stacking each stone, they rebuilt the structure together. Together they ruefully observed their carefully laid mortar trickle from the joints.

This episode demonstrates Fran's way of learning. Sure that they had bought worthless cement, she decided: "I'll go to town for a new supply; you're too polite to argue, Hammy." The patient dealer explained: the water-washed sand they had used was to blame. "Go over to the moraine, on Highway O. The sand over there'll do the job." He gave her a large bunch of bags. She drove to the moraine and filled two bags, each too heavy for her to lift. Only as she redistributed the sand into many bags did she understand his largesse. She loaded the truck, drove home, and mixed the mortar as Hammy relaid the stones. Then she rebuilt the weathered chimney, all by herself.[3] She had learned about water-washed sand and also to listen with fewer preconceptions.

Many saw Hammy making the decisions. Mrs. Errington noted Fran leaving a party directly upon his signal. Fran agreed: "Frederick had given me the look. It was time to go home." He set the boundaries. When, in the mid-1930s, he heard that a Necedah neighbor, thinking him a warden, planned to shoot him, he determined to set the matter straight. Game wardens had been shot in that county. Fran begged to go with him; women, she said, were a softening influence.

"I have made up my mind to go alone."

"But I'll wear a skirt—my green tweed skirt."

"Fran, the matter is settled. I'm going alone and on foot." No argument moved him. Next morning an anxious young wife watched her husband through binoculars from an upstairs window. He walked away, swinging his arms to show he wasn't carrying a gun. He returned triumphant: the story that he had heard, simply put, was a lie.[4] Their son was explicit: "Hammy made the big decisions. Because Fran talked more and was the active social planner, many saw her as dominant. However she might agitate for an activity, Hammy had the final word."

Hammy built on her preferences and skills. Her diligent efforts to please made it easy to listen to her opinions and to include her in activities. She so loved fieldwork that, to prevent any precautionary limiting of that activity in the last month of her ten-month first pregnancy, he carried dental floss with him in case the baby came in the field. Besides, since "She can out-walk, out-hunt, and even out-see most men," Hammy was able to concentrate on planning, recording, and organizing. Her formidable energy made her the logical one to hunt with outdoorsman Bill Longenecker, then director of the university arboretum in Madison. He never forgot that day. "She was eight months pregnant. We pushed through brush, waded ditches, climbed fences, pulled ourselves through mud. I have never been so tired in my life. When we finally got back to the house, there was Hammy, calmly typing away while she got supper."[5] She loved climbing trees; he made opportunities for her to do so, and on her sixtieth birthday, he presented her with her first pair of climbing irons.

Such stamina explains old friend and colleague Jim Hale's perceptive comment: "She was the engine, and he took very good care of the engine." Hammy spared her finances, taxes, and tasks like outhouse maintenance and tending the fires, and Fran, who knew how frustrating he found it to make repairs, would say to the gabboons should minor upkeep be called for, "Don't tell Hammy! I'll take it in."

That the Hamerstroms were also handsome, colorful, and nonconformist added appeal—as did their habit of performing. Fran's version of an outing with an Iowa game warden is a mini-drama. Startled by her appearance in the field, he made the mistake of asking Hammy, "Is she coming too?"

"Yes." But within minutes he turned to her. "Fran, see if you can see quail sign from the vantage point of that oak." She shinnied up the twenty-foot trunk, only to report that none was to be seen. They slogged on. Hours later, when Hammy called for a lunch break, the burly warden recommended a pleasant cafe in town.

"We don't want to lose any time," announced Hammy, pulling a chocolate bar out of his pocket. He divided it into three, ate his bit, brushed his hands together. "Better get on," he said cheerily.

"Wait! I have a second course," said Fran. She produced a frozen rutabaga that she had dug, scraped off the gumbo mud, and cut—all with the same knife. And at four-thirty, when the warden called quitting time, she responded brightly, "But Mr. Updegraf, it isn't dark yet!" Hammy flicked her a look of instant appreciation, and they pushed on, lifting one heavy foot after another until they reached the car, scraped the mud from their boots a final time, and drove back to the hotel. Then, at the door, Hammy asked Fran, "Would you like to go dancing?"

"Yes!" He turned to Mr. Updegraf. "Perhaps you would like to join us?" Poor Mr. Updegraf! He had received "the treatment." Meanwhile they had accomplished their survey, entertaining themselves in the process. Hammy, however, added a detail that she might not have: they both went immediately to bed and slept for twelve hours straight.

As Fran gleefully told such stories her repertoire grew. She was the star of many of her tales, and of almost all stories told by others. She chased and caught a wounded blue heron on Highway 51—in heels and formal dress. When she stripped to her underwear on a blustery April morning and waded a drainage ditch to shinny up a tree on the other side, she returned "all bark bloodied!" One observer of this feat called her an exhibitionist. Such tales, outrageous or entertaining, exaggerated or factual, were told and retold by friends, visitors, acquaintances, and critics. They often reveal more of the values of the teller than of Hamerstrom realities.

For example a fellow biologist, a trained observer, spoke of their visit to Minneapolis. His first comment was, "They each took a hot bath." His account, liberally punctuated with exclamations and chuckles, went on to describe a very pregnant Fran's attire: torn, tattered jeans and a ragged shirt, inappropriate garb for a visit to a university museum. And, "She ran up three flights of stairs!" he breathlessly reported. His most relished story was of joining Hammy, Fran, and first-born Alan in a sandhill crane watch, miles down an old logging road in Necedah. After an hour or so he remembered. "Where's the baby?"

"Back at the car."

An hour's walk brought them to a bare baby boy, "tan as a coffee bean," gurgling and cooing on the ground (as he specifically said) by the car. "He must have been there several hours. They were both quite unconcerned."

A baby on the ground would be vulnerable to any hungry predator. Elva shook her head when she heard the tale. "He might have been bare, but on the ground? Never. Coyotes could have gotten him. They always put my basket on top of the car!"

Although a surprisingly frequent response to queries about them was a question: "Why didn't they at least put in a *bathroom*," close friends and coworkers understood the necessity of their choices. Gustav Kramer's wife Neni, as a very old lady, had vivid memories of a 1955 visit. "Their house was very . . . " she hesitated, searched for the right word, "primitive." Primitive meant practical for Fran and Hammy. Home and vehicles were multipurpose; essential tools were right at hand in the storeroom. Books stood logically on shelves built to measure by Hammy. They had the important things: an electric stove, a pressure cooker, a refrigerator, and a freezer. The outhouse and the tin tub—or a dip in the pond—took the place of an expensive bathroom. Visitors, troubled by the idea of free-flying owls, or disgusted when they found odorous mice cages in the house or a dead cat or snake in the refrigerator, might react with shock and distaste. Elva remembers a well-kept home: the occasional disorder was not the rule.

They searched constantly for household help.[6] Mrs. Chamberlain did the laundry until visiting Posey exclaimed, "Two dollars! That's not enough! Here's $5." That quick price rise, Fran claimed, returned her to the washboard until the blessed advent of Laundromats.[7]

When there wasn't help, benign neglect would do. A would-be helpful German visitor offered advice. "I pick up daily," she volunteered. "Tuesday I wash, Wednesday I iron, and if I clean on Thursday, I finish by noon! Then all I have to do on weekends is cook." Fran considered this thoughtfully, rejecting the list of activities she might have offered up in return. "I clean when it's dirty," she responded. Hammy, often part of the family cleaning marathons, could not repress a smile.[8]

Fran caught and tended the mice used as bait for trapping; she did most of the gardening; she chose to replace nails in the red tarpaper on the root-cellar roof on a very hot summer day while Hammy worked in his spacious cool office. "He could put his hand on anything he wanted among all those piles and shelves," said an admiring gabboon. He paid bills, did the income tax, wrote quarterly reports, and supervised statewide activities. He, in immaculate shorts and shirt, almost always mowed the lawn, for only he would be sure to go around the patches of wildflowers in the yard. He did inside chores as necessary, sweeping up the woodchips and dust inevitable to wood heat; sometimes doing dishes, pumping water, and broiling cheese sandwiches for the observers' second breakfasts. He simplified such tasks: a young woman, reaching for a cup to dry was startled by his peremptory command: "Don't move that cup! That's where I prop plates to dry." Their disregard of conventional roles marked them in an area where most women said with smug satisfaction, "He does the outside work, *my* work is inside."

Years later, a reflective Fran said, "Hammy always knew who he was. That isn't true of me. I've always had to prove myself." And she did so, over and over. Remembering her efficiency during the Action, he set her to managing arrangements for placing observers in blinds. She handled the address file, reservations, and complications. She made sure to assign important contributors to their one luxurious wooden blind. She welcomed reporters, cultivated useful contacts, entertained constantly, and saved every clipping.

They hosted informal but purposeful seminars when the departmental research bureau gathered to help with the booming. The subject matter, as one would expect, was conservation—but details are lost. Instead, informants remember the odd occurrence. Ralph Hopkins, former head of the conservation department's Wautoma district, saw their big rooster come up and spike Ruth Hine in the ankle. "I kept my eye on it," Hopkins said, "but when my attention wavered, by gum, it sneaked up and spiked me. I turned, took aim, and drop kicked that chicken as far as that window! It didn't spike anyone else that day." Between sessions, Fran cooked food for the gang.

"Partner," did not mean helpmate to this couple. Fran watched Hazel Grange serve Wallace. She did not approve of what she saw. "She was his slave—she'd drop everything, cut wood, feed the stoves, or rush off to northern Minnesota to trap snowshoe hares from an unheated shanty. She kept the books on the venison he sold to fancy New York restaurants. He would shout: 'Time to fight fire!' and out came Hazel, with a shovel and wet gunnysack. That pet fox and captive sandhill certainly weren't *her* idea."

The pets in the Hamerstrom home were unquestionably Fran's idea, though she maintained that Hammy enjoyed them. "No, he never wanted to get rid of them. We rigged the net because the owls, fascinated by his typing, would watch his fingers move and suddenly attack. He laughed at our short-eared owl when it hissed and glared as he came near. He suggested the name "Tantrum," you know. Perfect!"

The long, screened front porch was rarely without an avian occupant. Hammy was pleased with her help to a man who phoned her about a frantic hawk trapped in the potato-seed storage warehouse. He marveled at her. "She came right over and, d'you know, she *talked* to that hawk. Before long she had it."

"They'll remember that," Hammy said, in quiet satisfaction. "Let's hope it makes the rounds as the knitting story did."

Neither of them knew a forty-hour week; Saturdays and Sundays were like any other day. In slow seasons they took time for fishing or hunting, and vacations were never for humdrum tasks like putting up storm windows or rebuilding sheds. They might hunt, or trap water shrews in northern Wisconsin, or head to Langruth, Manitoba, to camp, observe, and trap during the fall hawk migration. Learning in situations of adventure or solitude was their idea of a refreshing change. Their mutual disregard of nine-to-five routines and dress-for-success conventions might seem quite appropriate today, but then it raised more than eyebrows.

The necessary parsimony of student days became a habit. They saved paper for reuse long before recycling was common; Fran wrote all her books on used paper. They looked ahead: in 1943, embarking on a joint career, they arranged mutual power of attorney—hardly a customary step for a young couple of the times.

They accepted each other's frailties and peculiarities. He endured kitchen problems: Fran, busy to distraction, would put the coffee on and let it boil while putting nylon nooses on a trap or feeding her captive heron. "FRAN!" would come a stentorian voice from the office. "Coffee's boiling!" Or, "Frying pan's too hot!"

Hammy was meticulous in every task. The apple slices he produced for pies were exactly even. Larry Crawford wondered at it: "He was careful of everything. I stored my hay in their barn. Once he phoned to ask me to come look at it. He was worried; the roof had a little bit of a leak. 'I'll get it fixed right away,' he assured me." When he made a skin it took him half-a-day to skin, treat, and stuff the specimen. He finished by stowing it carefully in a nylon stocking to smooth and round the feathers and did more arranging the next day. Fran wasn't that particular. She said: "I do a skin in twenty minutes—his skins were works of art." Each respected the other's style.[9]

Hammy preparing sharp-tail grouse skins at a hunting camp in the early 1950s.
Fran called Hammy's skins "works of art."

A stickler for detail, he could be hard to please. Gabboon Mark
Kopeny remembered trying to repair vandalized windows on a marsh
shack they used as a research base. "He wanted the right tools, and the
right skills. Glazing windows with us—who had neither—tried his
patience." Pete Frost did house repairs for them. "We had to take the
vines down, lay them gently on the ground and protect them. Then we
had to put up the siding—used material, so it would match—and put
the vines back up as they had been. When we put new tar paper on the
shed roof, it had to be red—because it had been red, y'know."

Hammy looked at every question from a long-range viewpoint, ne-
glecting no detail. He provided helpers with maps drawn to scale and
labeled to distinguish a track from a farm road or a footpath. He described the vegetation; warned of dangerous or over-affectionate dogs;
characterized the farmer as friendly, taciturn, or inquisitive. "Get
permission from the landowner, no matter what the errand," he said.
"Write everything down, even if it is a one-time occurrence! Use a pencil; ink runs if the paper gets wet." At task's end, he wanted a complete

report of trouble or discoveries. Thus he developed a mental file of variables so that as tactician, he could set standards, determine the scope of tasks, plan schedules, the meeting places, and then the specific priorities. One helper looked back, "I vividly recall the exhaustive planning sessions prior to going into the field to check traps. I couldn't appreciate the need for such detailed discussion of a straightforward field mission. Later, directing field crews of my own, I realized how important it was."[10]

In the early Plainfield days Hammy still smoked, and it showed in his demeanor. "I felt kind of sorry for Fran then," reported Dan Berger. Both Hamerstroms had announced their intention to quit smoking to Mary Mattson; he succeeded, Fran could not. Hammy became ever more patient and composed, but Fran remained her impetuous self. Berger, who worked closely with them for years, "longed for that early firmness" when he experienced her immoderate enthusiasm about trapping at Cedar Grove trapping station.[11] Another coworker said, "He'd call her on any deviation. He had to." He checked her work as he did everyone's. They must be sure of the dates of the peak numbers of hens on booming grounds. "Fran, did you check the hen peaks? Here's my worksheet."

He had an encyclopedic mind. A colleague remembered it.

Walking along, he'd stop, identify a forb, (he always used that term, never "plant") point out an anomaly. I never saw him stumped. "Now look at this," he'd suddenly say. Then you'd get a dissertation, complete with Latin names. We were sitting at the table in the old house with hawks flying around and two prairie chickens lying on the floor in cases, waiting to be weighed. He was talking, and as he spoke he suddenly put his hand to his temple, picked off a minute insect of some kind, regarded it dispassionately. "Ah!" he said, "There's a *phebotomus*!" He got up, went to the office cabinet by the door, got a glass vial, put it in, corked it, stowed it, and resumed his conversation. He didn't lose a beat.

Another companion once wondered about a weed he did not recognize; Hammy identified it, but six weeks later he came back with a correction of the subspecies name he had used.[12]

Mutual support characterized them. Fran protected his precious time and their time together. Should someone come unequipped with boots, jackets, or binoculars she fitted them out from their informal loan closet. She devised errands for early arrivals. She might say, "Do go check out the woodcock peenting; there are some in the back field. Be back at 7:30."[13] Or she would beg them to rush back into Plainfield and buy four dozen eggs. "I'm short of them for breakfast. Briefing starts promptly at 7:45." In turn, Hammy would lay down the law to her helpers. "From now on no more snooping near the spray people's airstrip. I wouldn't want to have to explain to your families how you happened to get shot."[14]

Were they competitive? Yes, in a teasing way. Hammy would bring Fran down to earth. "You know, when I made my first big sale, to *National Wildlife,* I looked at the check and figured the pay for each word. I rushed in to his study. 'Hammy! I'm getting sixty-seven cents a word!' The magazine had used a photograph he had submitted. He looked at me in that imperturbable way, and responded dryly, 'I got $50 for 1/125th of a second.'"

Patient, phlegmatic, tolerant, he was the flywheel to her quick-tempered, reactive personality. Yet her frequent use of the phrase, "I said to Hammy . . ." indicates that she generated many ideas and was the sounding board for many more. Her assessment of her influence on him was clear and unequivocal. "Hammy could have been a stuffy scientist. I remembered Leopold's maxim: 'You've got to reach people,' and I helped him do it."

Of such interaction was this team forged. Their mutual admiration was public and evident. It magnified their effectiveness. They had to be effective for they had an enormous task, and one that too few people understood. Helping the public understand was the early purpose of Fran's story-telling mode. Her audience of booming guests gave her a public she could instruct while entertaining. (Once, when Hammy mildly interjected a correction, Cleveland Grant, a photographer, chided him. "Never spoil a good story for the sake of the truth!")

Alan feared such inaccuracy would damage his parents' standing. Should he attempt to tone his mother down, however, Hammy chided him. "You'll only make her unhappy," he said. Alan chafed under this advice, but when he drove her home from Texas the first spring after Hammy's death, he saw the genuine pleasure friends took in her presence and her tales. He wrote her, confessing that his early worry was unnecessary; her flamboyant style and nonconformist approach created the happiness that he too, wanted for her. Poignantly, he continued, "I am fifty and you are eighty-three; you have legions of fond admirers, and I, the respect of a limited cohort of people who really do not know me very well."[15]

Strangers came to see them: young people who had heard Jean Ferraca interviews on WHA, the Wisconsin public radio outlet, would-be-biologists seeking introductions, hunters, bird-watchers, readers, and the just plain curious. They influenced people in a range of ways, some inconsequential: "They have a Volkswagen bus. When I can afford a car, I intend to get a Volkswagen bus." Some were profound: a college friend visiting for a weekend with Elva, went back to Madison, dropped out of her humdrum course of study, to return only after a season of work and travel. Later she told Elva, "Your mother changed my life! When I went back to college I knew what I wanted, and I got a lot out of it."

It is difficult to assess such influence, but clearly, a kind of mystique about them invited imitation. Visitors passed on news of this unique couple who didn't bicker, never turned away an interested guest, and demonstrated an inspiring way of life that those who knew them wished somehow to emulate.

The legend invites reflection: Was their collaboration relevant to the larger field of scientific investigation? Comparison to other scientific couples led to the question of sex roles and science: Was this team a meld of feminine and masculine approaches to research? Obviously, there are men researchers who look at their subjects in ways that those careless of connotation call "feminine." A single telling example is that of Roger Fouts. The chronicle of his devoted work with Washoe, the chimpanzee who became proficient in sign language, is told in *Next of Kin*.[16] That book records, in painstaking detail, Fouts's meticulous observation and method, plus a compassionate approach that culminated in his creation of a chimpanzee refuge. Is the observation that I often heard that Fran was the more imaginative and subjective partner, Hammy (the scientist) the more rigorous, exacting, and objective one, merited?

Evelyn Fox Keller, brings a thoughtful distinction to such a question in her work about Nobel Prize–winner Barbara McClintock's life.[17] Since objectivity is usually seen and defined in our culture as "masculine" and subjectivity as "feminine," and since various value judgments rise from those views, Keller defines two kinds of objectivity. A "dynamic objectivity," grants the world and its objects an independent integrity while recognizing our human connections to that world in a mode of empathy. "Static objectivity," in contrast, pursues knowledge by severing a subject from oneself and then dealing with it as an object necessarily removed from the observer. The first mode is akin to love; the second can lead to domination, and in her view, oversight and distortion.

Leopold's eloquent words speak of the results of domination to the obverse: "Do we not already sing our love for . . . the land . . . ? Yes, but just what and whom do we love? Certainly not the soil which we are sending helter-skelter down river. Certainly not the waters, which we assume have no function except to turn turbines, float barges, and carry off sewage. Certainly not the animals, of which we have already extirpated many of the largest and most beautiful of species."[18] Would he not support Keller's calls for a gender-free science that will encourage and reward dynamic objectivity?

Did the Hamerstroms practice dynamic objectivity? A memory of something a Plainfield farmer once said came to mind. That man, whose wife had, for a time, done the Hamerstrom's laundry, testified: "I know them—*and they just love the prairie chickens.*" Simone Weil defined love as the giving of attention to an object. I remembered how Fran recalled Hammy's faithful attention to the land when she saw an entrepreneur's bulldozer piling the woods bordering their farm into ruined heaps for burning. "I'm glad he didn't live to see this," she said. Fran's care of her eagles, Chrys and Nancy, demonstrated meticulous attention. Clearly, each of these scientists loved what they observed. Their efforts were to preserve a species; their hunting was a modest ritual harvesting of surplus stock. (There was no sign of a trophy in their home.) They lived in harmony with their acres as well as with their hawks, owls, and other subjects.

Fran insisted that children must have wild creatures as pets, because they taught their owners so much.[19] Consider *An Eagle to the Sky.* The reader sees—and feels—Fran's way with her eagles. When an observer called her first eagle "captive," she considered the term.

"In a way, I suppose she was as a dog or . . . one's own children are captive: one influences them. I spent an enormous amount of time with my eagle, and there are those who would inquire, 'Which is the captive?'"

She also said: "I was learning to learn from an eagle. . . . I saw that satisfactory motion . . . the luxurious rocking from side to side as she settled onto her own egg." She joined in the bird's activities: "We worked on the nest an hour or two a day, which seemed to satisfy her." Later, "I realized that it was my duty to relieve Chrys at the nest, [so] I took my turn day after day."[20] In that task of keeping the egg warm, her behavior was, perhaps, excessively maternal. She tended the eggs for Chrys in all weather, where a male researcher might surely—and more easily and safely—have resorted to an incubator.

Aldo Leopold reminded us "that men are only fellow-voyagers with other creatures in the odyssey of evolution. This new knowledge should have given us, by this time, a sense of kinship with fellow creatures; a wish to live and let live; a sense of wonder over the magnitude and duration of the biotic enterprise."[21] He encouraged amateurs of nature to cherish and restore the land they owned. Like him, Hammy and Fran sought to live in harmony with the land. Without their ever having heard the term, I suggest that each of these three scientists practiced dynamic objectivism.

Other contemporary couples, Wallace and Hazel Grange and Olaus and Mardy Murie were, like the Hamerstroms and at about the same time, engaged with wildlife. Hazel Grange's *Live Arrival Guaranteed: A Sandhill Memoir* is her captivating account of the complexities and perils of being a conservationist in the 1930s and 1940s in the mode of a detailed personal journal.

The Granges kept alive a farm venture in Door County as well as the game farm that they founded in Wood County during the years when Wallace was earning a living in various jobs with the U.S. Biological Survey in Washington and the Wisconsin Department of Natural Resources. They survived depression, several house fires, drought, community opposition, almost fatal lightning strikes, market variations in fur prices, and a venal game commission. Hazel was the 100-percent-loyal wife, holding things together in her husband's absences. Although she did not fully understand his passion for natural propagation of wildlife, she cooked for the crews, fed animals, kept house with rudimentary utilities, and managed harvests and help. Her internal code forbade any complaint. Happily she came to understand their labors; Wallace read the tribute she wrote in her memoir before he died in 1968. "Never anywhere on earth, so far as I know, had anyone else conceived or brought into being a community of wildlife adjusted to its basic requirements and population capabilities on a pay-as-you-go plan from its own surplus. . . . Now for the first time, I really understood what had for so long bewildered me. . . . It was far more than 'work.' It was artistry."[22]

Wallace's creative husbandry of his land, like Hammy's, had to overcome strong public opposition. Both operated under Leopold's hope that private landowners would practice conservation; yet in the end, public stewardship became necessary for both of their major enterprises. But Hazel was, necessarily and understandably, more of a cooperator with her husband than his collaborator.

A collaborator is an intimate partner who deeply shares the ideals and goals of the coworker as well as the work. Fran and Hammy's pattern of shared work and especially of writing together—where they served as trusted, thoughtful, and consistent critics of each other's copy—indicates deep and subtle understandings between them. That closeness may have risen from the alienation from their communities of origin because of their unconventional marriage, as Vera John-Steiner speculates about another collaborative couple, Will and Ariel Durant.[23]

Fran had experienced isolation and its effects, as in Necedah where they answered visitors' questions together, even using the very same words. Simone de Beauvoir described her collaboration with Sartre in like phrases: "We always criticized, corrected, and ratified each other's thoughts. . . . we might almost be said to think in common . . . our attempts to grasp the world are undertaken with the same tools . . . guided by the same touchstones . . . we have been known to produce identical answers."[24]

Margaret Murie started as Olaus's assistant, learning species, flora, and method as she helped him. Olaus and Hammy were both calm, serious biologists. Deliberate, introverted Mardy appears to have been the absolute opposite of Fran, but she, too, was a total, loving partner who complimented her husband's personality. Mardy moved beyond assistant to political activist and influential author; her enchanting memoir of her early life and marriage, *Two in the Far North,* has enjoyed deserved and considerable success. Her role in pushing the Alaska Wildlife Refuge act through Congress won her accolades to which Fran, in a less remarkable arena, could hardly aspire.

It seems significant to me that both couples shared a love for the "innocent beauty" of the wild and a kinship with the untempered land. Both wanted to preserve wildness. Each wife was prepared by an unusual childhood for her future role: Fran's early rebellions led her to her fresh and wholehearted interest in wild things; Mardy's childhood in Alaska gave her the same curiosity and need for romantic adventure that so motivated Fran. Both inspired younger collaborators, both determined that they always wanted to "be together" with their partners and managed their lives to make that outcome possible. Both, interestingly, loved to dance.

The natural sciences were, for the Hamerstroms, fertile ground for close collaboration, but their decision not to let the antinepotism rules, then common at many universities, blight her desire to do research was fundamental. It allowed them to do what they wanted to do in a sustained relationship. Other scientists, including Olaus Murie—who left the Biological Survey to work directly for the environment in the Wilderness Society—chose personal satisfaction over security and status.

*Creative Couples in the Sciences,* a book that explores scientific partnerships over the later years of the nineteenth century and much of the twentieth, provides a useful common vocabulary, a set of criteria, and short analytical biographies of nine partnerships of various depths and complexity.[25] Four unusually rich and intimate partnerships, three of which were in the natural sciences, are germane.

A "partnership of equals" often includes lifelong collaboration, unusually close companionship, a practice of sharing ideas and skills, and often, a career pattern of "taking turns." The male partner in each pair cited was a husband who did his part; the wife had confidence and self-assurance, often derived from an unusual childhood. Each career required considerable dedication and sacrifice; several couples undertook public service or a reform activity, as Fran and Hammy did. Finally, these partnering lives often featured a certain amount of unconventionality. What is most refreshing to see is the considerable amount of mutual joy among them.

Three of the husbands supported their wives at some cost to their own careers: At sixty, physicist Thomas Lonsdale, who had been the primary breadwinner, retired to take over the huge correspondence generated by crystallographer Kathleen Lonsdale's work on peace, disarmament, and prison reform so that she could continue with her scientific career. Cyril Berkeley, originally a bacteriologist, gave up most of his own research to help his wife with her polychaete taxonomy after the 1930s and well into the 1960s. Astronomer Frank Hogg, after years of being helped by his wife, served as astronomer Helen Hogg's informal assistant while she worked on a telescope that required two people to manipulate. (She was the first person to use the telescope for direct photography.) Frieda Blanchard was a botanist, Frank, a zoologist. They collaborated by talking over ideas, keeping records together, and helping each other with the eternal task of revising and proofreading papers. She did drawings and illustrations for him and accompanied him on his collecting trips. One year-long trip required that they leave one child with a cooperative relative and take their infant (in a sling she made) into the forests of Tasmania and Australia. That trip was a high point in their lives.

Fred and Frances Hamerstrom were equally creative collaborators, not just in their work. They were also models of a seamless mode of living and working in a gender-free science. Were they simply fortunate or atypical? They undoubtedly represent a gentler science—the earlier investigative and observational mode that preceded World War II again connects them to their mentor, Aldo Leopold. He deplored what he called "Power Science," science dependent on large government grants, industrial support for the land-grant universities, the National Science Foundation, and research aimed at profiting from rather than preserving the land.[26] Whether that science has limited or enlarged potentials for collaboration remains to be seen.

In a recent lecture sponsored by the Santa Fe Institute, Vera John-Steiner suggested that there is growth in collaborative opportunities: coauthoring, the institutions and tools that are developing to make collaboration easier, and the scientific "thought communities," now flourishing in a culture that has previously undervalued collaborative efforts.[27] She cited the richness that comes from the shared creations of investigators from different disciplines, and emphasized that mutual trust allows multiple perspectives to create new concepts out of productive conflict and shared delight.

The concepts in her book recall Hamerstrom practice. Their work with the gabboons has resulted in a loosely linked community of environmentally sensitive and productive people that suggest a "thought community." Those who consider the Hamerstroms as mere "mid-level biologists" might consider the depth and breadth of their influence on a worldwide range of people. Their connections, their regular contributions to international conferences, and their helping young people to be ready to work in the field from Canada to the Orkneys to Saudi Arabia and Africa has contributed to the growth of the environmental movement. The Hamerstrom values, reflected in their admirers, may well influence our future.

# 18

# Death of a Biologist

*The deepest feeling always shows itself in silence; not silence but restraint.*

—Marianne Moore

We missed seeing Fran and Hammy the summer of 1989, and they went to Guatemala that fall. I called them after Thanksgiving. "Fran! do come to dinner Saturday—we haven't seen you for ages." There was a long pause.

"Hammy's had the punies, Helen, ever since we came back from Guatemala. Why don't you ask him?" It was odd; Fran always made the social arrangements. He came to the phone. Yes, they'd like to come and leftover turkey sounded fine. He admired the special bottle of wine we opened, and we spent a tranquil evening by the fireplace. He mentioned his illness—something tropical, perhaps, picked up on their trip.

Hammy knew my brother Malcolm, a physician in Houston. "Let me give you Malcolm's phone number, Hammy. If you need a consultation, he would know someone good down there."

"Thanks, Helen, our doctor is checking a couple of things out. I'll call you if I need it."

Monday afternoon, the phone rang. It was Hammy. "I'd like Malcolm's number from you, Helen. I just got back from our doctor in Rapids. There's a ten centimeter growth on my pancreas."

"Oh, Hammy, that is bad news!"

"We'll be going on to Texas as we always do, and I'll see a specialist there."

"Here's the number . . . and his office number. Call if we can help in any way."

When I told Kip he said, "Let's go right up." We found Fran typing in her office, Hammy at the oval dining table, sorting the copious mail. He stood to greet me; we hugged. Kip shook his hand. "Hammy, I don't like your news one bit."

He looked at him dispassionately. "Hell, Kip," he said, "something like this was bound to happen. I'm eighty, after all." With masses of work to clear up, they couldn't leave for about two weeks, but Hammy said he would see the recommended Houston oncologist before going on to Welder. Fran came in, offered coffee or wine. Hammy made one thing clear. "We don't want you to speak of this to anyone—I mean anyone!" he said, severely. "We don't need gloomy faces around here."

I couldn't get them out of my mind. Their winter disappearance would mean, this time, that people who loved them might not see him again. I called her. "Don't say no right away, Fran. We haven't had a party in ages; let me have one for you before you go south. Talk it over and call me. Just give me a list to invite."

She accepted, with a list—friends, colleagues, a favorite newspaperman, a dean—and instructions that were pure Fran. "Tell the men that it would make their wives very happy to wear long dresses." They needed to leave in two weeks, but the Friday of the weekend before their departure was free.

The day of the party she asked if she could bring visitors, and did so. Dale and Elva were also in attendance. Regal in a caftan, Fran was very much herself. Hammy held quiet court from a big chair by the fire. I brought him a special tidbit—small cubes of firm tofu in a hot, gingered meat sauce. "Hammy! I think you'll like this." He nibbled obediently. "Guess what it is." He gave me a knowing look. "Tofu!" he growled, then threw back his head, and laughed, remembering that long-ago day in Minneapolis when the owlets arrived. He said goodbye with a hug, and warm thanks.

To our surprise, we found Fran and Hammy in the concourse of a crowded mall in Stevens Point the next morning, selling her books from a long table. Hammy took money, chatted with friends who stopped by. Fran manned the table.[1] It was business as usual.

When Dale and Elva, who drove the Hamerstrom car to Texas, got to Houston Hammy met them. "The cancer is far advanced," he said, "but Fran and I will go to Welder after Christmas as we intended."

The dean of prairie grouse studies in his seventies, in 1981.

Later, Alan and Helen shared that last Christmas with his parents in a beachside motel. Alan reported, "There was little talk of plans and options—that was clearly what they wanted."

At Welder the staff provided the privacy they cherished. The Hamerstroms did appear at various functions and met the current group of graduate students. They asked my brother Malcolm and his wife, Ursula, to join them for a weekend; the four of them shared dinner with the assembled residents in the "roundhouse," and Hammy spoke briefly but insightfully.

Next evening they drove the Volkswagen, complete with caged mice and traps, to the creek where shore birds used to congregate before the current drought, setting a trap out for the hawk they saw. "We'll check it on the way back." Ursula spotted a bobcat—not the usual tawny hue, but "almost charcoal." Hammy spoke approvingly "You have a marvelous eye. You really should do more with it."

At the dry creek silhouettes of gaunt trees killed by the drought stood black against the sunset. Conversing turkey vultures burdened their branches. "It was eerie," Ursula recalls. Driving back, Hammy alerted them: "Look closely, a family of javelinas often crosses here." He slowed as a group of the wild pigs disappeared into the brush. They retrieved the bal-chatri and recaged the mouse.

My skeptical brother called me after that "wonderful weekend." "They were held in obvious reverence," he said. "Hammy's observations were deliberate. He struck me as accurate to a point well beyond most men's reach." Malcolm noted the breadth of his reading, his knowledge of general medicine, literature, and world affairs. He told me that it was unlikely that Hammy would ever return to Wisconsin.

Some days later a group of students from the school of natural resources in Stevens Point arrived, well past their scheduled arrival time; the faculty leader called from Sinton to arrange for opening the locked gate. On their arrival, "There was Fred. He drove up to let us in. I didn't think much about it at the time, but later, I remembered. It was a kindly thing for him to do. He was very much himself—helpful, welcoming, pleasant."

The Hamerstroms didn't approve of prolonging life. Fran explored options and decided that Oregon, Elva's home, had the most benign environment for terminal illness. Fran drove most of the 2,300 miles to Roseburg while he drowsed beside her, waking to manage the "difficult bits around cities, with all those signs."

Elva found them a cottage some twenty-five miles from Roseburg. "I found a wild place with conveniences where they would be happiest—by themselves. Hospice nurses came; we visited often and did errands for them as requested. Fran walked to the tiny post office nearby for mail each day; the remaining hours went to writing her autobiography. Hammy edited each chapter as she finished it. 'We're getting quite a lot done,' she would report proudly. Hammy dressed for meals, even when that meant simply putting on a sports coat over the handsome pajamas Fran had made for him."

Elva's birthday was 25 March. "Hammy didn't want to mar, or mark, that day." He asked Fran, "How long do you think it will take to finish?"

"I have three chapters. I think I can do it in three days."

"I don't know if I can hold out that long." The editing seemed to energize him. She wrote the last two chapters in a day; he roused himself to edit them and to complete the introduction on the day before he died, as Fran reported.

"Was that so?" I asked Elva.

"I expect that they talked out a sketch of it that she roughed out on the typewriter. He may have added that precise last sentence; I saw him with that page. They had collaborated so often, it was hard to tell who wrote what."

Here is that unused introduction. (*A Double Life* was published without it four years later.)

For almost 60 years, Fran has regaled me with stories of her childhood—vivid stories of long ago.

One day, when I came in from setting traps, I asked, "How is the writing coming?"

Fran . . . whom I'd seen cry only twice during the 58 years of our married life—hurled herself into my arms and burst into uncontrollable tears.

"Good God! What now?"

She was writing a chapter—as it happens the chapter about not being allowed to accept a bowl of goldfish. The unfairness and the misery of that refusal!

An old cook who understood her love for animals, had come to the front door and brought "a present for Miss Frances."

But the grownups said no and made her give them back.

Fran didn't dare cry then, she was so thoroughly under the rule of these powerful adults. Writing the chapter had brought back the whole story. Finally, she cried. Seventy-four years after that cruel dictate she was sobbing in my arms. The vividness of her emotional recall comes through over and over again in her writing.

Alan wrote, asking if he should fly out. Hammy said he needn't. "Fran and I want to spend the time together." They wanted no service. "Funerals are ghoulish; we never go." An obituary was appropriate; Fran dialed Bill Janz, a writer for the *Milwaukee Sentinel* whose work had pleased them, and gave the phone to Hammy. "I've got cancer, Bill, and it won't be long. But," he continued: "I am in a beautiful place. The Umpqua river runs right beside our window. I can watch it and listen to it. I can see maples just budding out, Douglas fir, and red alder. You know, Bill, life has given me adventure, public service and a loving relationship with my wife. I couldn't have asked for more—And death is not unexpected, now that we are in our eighties. It is something we are doing together. Thank you for doing this for me." He put down the phone. Shortly, he turned to Fran. "Fran, I'd like you to call Bill back. Tell him about the memorial[2]—and I didn't tell him how minuscule the moment of death seems compared to our sixty-one years together."

Three days later Fran wakened Janz with a phone call at 3:00 A.M. He could now send the obituary out.[3] Fran's announcement of his death—when they hadn't even known he was ill—startled family members. His brother understood. "He was always reticent. Recently, I mentioned to him that I had had peritonitis. 'I had it too! a year ago! Interesting coincidence.'" That was Hammy's nod to the time when a ruptured appendix on the way back from a family wedding caused a rushed return to a Boston hospital from New York. Fran made him go back. "I saw the pain he was in; I saved his life."

Ten days after his death, Fran drove back to Wisconsin alone, refusing Alan's offer to drive back with her. She had made other long, solo drives.

Alan came to closure: "It was almost as if Hammy was away on a trip. He was often out of our lives for periods of years anyway; we got along with that. He made it as easy for me as it could possibly be, so I ended up with a good recollection."[4]

Fran spoke of good memories. "One hospice nurse came to see us every day. She didn't need to, she just wanted to." And, her face shining, "He called me his delight." But grieving! Sympathy! "Everybody wants to hug me!" she complained. "Even people I don't know!" She burned most of the condolence letters. Her mourning was inward.

She did keep a watchful eye on what was said. She carefully reviewed the appreciative four-page "In Memoriam" that John Emlen sent her before publication and made six suggestions, tactfully explaining each one in a vivid phrase. She concluded with praise: "Johnny—It's just perfect." Another obituary was more troubling. It said, "Regrettably, Fred did not seem to enjoy writing. . . . His list of publications was adequate but not extensive."[5] Adequate! Outraged, she wrote the offending author, who pointed out that he had called Hammy "the ideal role model for anyone with aspirations to be professionally involved with the welfare of our wildlife resource." She moved on with her work. "Much grieving would be selfish. One should go ahead and do what's right."

Friends put Hammy in a special category. Graduate school comrades remembered a weekend of "bucking for exams and drinking my father's Irish potato whiskey. I can remember Hammy this minute talking in his slow, deliberate way. He was so humorous, so succinct. A darling of a man."

"There isn't a person, anywhere, that I would rather have had for a friend than Fred Hamerstrom. He was the best there was," said Thad Smith. Ginny Emlen remembered him as "always courtly. One time I said, 'I feel inferior; I'm not a biologist.' I'll never forget his answer. 'But Ginny, you are doing something much more important. You make people happy.'" Many remembered his attentive presence. Only recently a member of the Tympanuchus society told me, "I can think of five or six people in my life who were outstanding, who made an indelible impression on me. The Hamerstroms were one of that group." Others honored his gentleness, magnanimity, dedication, patience, and consideration. He was seen as a pioneer, a world authority on grouse, a gifted editor and "midwife" of scientific papers, and as the engineer of the scatter pattern of conservation planning, now used in the management of other critical habitats around the world. Ray Anderson summed it all up with the felicitous ending phrase of a detailed and accurate tribute: "I felt I was in the presence of greatness."[6]

I loved Hammy and have been impressed by the unanimity of tributes to him. He was a man ahead of his time and glad to be where he was. I find myself wishing that I could talk with him about the latest grammatical monstrosity, or current pronouncements on ecology and conservation, or about the plight of the lesser prairie chicken. He would never have said, "I told you so." Secure in himself and in his achievements, he didn't need to claim credit or to gloat.

We all missed him. The great house seemed tilted without that steady presence. A young friend, one of the many that loved him, made the truthful, grave statement: "It is hard to imagine a world without Frederick Hamerstrom." Their much loved friend artist Jonathan Wilde put it this way: "His respect for nature was almost spiritual. They lived in harmony with their surroundings, not dominating them. They showed that you could be happy without running hot water or buying something every minute. And the way he took people as they were, without condescending—this was a person who was all of a piece."

Still, the record of his life, a veritable manual of the practice of landscape ecology in this complex world, remains. He taught ways of dealing with economic pressure and human shortsightedness, of banishing greed and living fully without overconsumption, of behaving—in the words of one friend—in a seamless way. He was indeed all of a piece. His legacy to the hundreds of people who saw him in action and worked with him may be impossible to quantify, but it is there, yeasting, working. His immortality lies in the living memory and echoing of his integrity, the creativity and reliability of his ideas, and the example of his loving, patient, and persistent practice.

# 19

# Fran

*I spit into the face of Time*
*That has transfigured me*

—William Butler Yeats

The Hamerstroms remained somewhat aloof from activities on the marsh after they left the DNR. Fran maintained that "It was simply too painful for us to abandon the project." Yet in 1990 she wrote to represent Hammy's concerns as she saw them to Jim Keir, the manager of the Buena Vista, Leola, and Paul Olson management areas. Keir had called a meeting with Ray Anderson, Jim Hale, Dick Hunt, Ron Westemeier, Ray Anderson, and some of the council of chiefs of the Society of Tympanuchus Cupido Pinnatus to discuss year-round management.

Fran's memo objected to excessive disturbance of the grassland. Cattle were grazing too early and too long. Fences, trees, old foundations, and thickets—essential for winter roosting—had been removed. Plowing sod for productive farm units limited permanent grassland. "The management plan," she fumed, "is 'by the numbers.'" Planning seventeen years in advance ignored likely changes. "No three-year droughts or disastrous fires, etc. for the next seventeen years! . . . I recommend built in flexibility." She felt deeply about preserving Hammy's legacy, and she was worried about the next population low.

Her real priorities were living abundantly and fulfilling her own long-term goals. "Now I can do the things Hammy wouldn't do. He hated the tropics—the damp, the mold, the fungus infections, and the parasites. I like hot weather and jungles." So she went to Zaire, where she hunted with the Pygmies, and then to Peru, where a guide took her deep into the jungle. The Gulf War delayed for a year her visit to a

290

former gabboon, then the well-paid falcon keeper for one of the sheiks. She went to Italy (where she ended up in the hospital) and visited the European children and a gabboon or two. She hoped to write another book, one about the reactions of a Pygmy, or of a rain-forest inhabitant to the North American boreal forest. She ended up remembering her childhood and distilling meaning from the exceptional life she had drawn from those unlikely roots.

Fran was full of contradictions. Sometimes autocratic and hasty, she could be measured and a considerate listener. She might seem self-absorbed, yet she loved to make people happy. She was alternately abrupt and thoughtful. She might smooth over a gaffe with gentle good humor, yet she often corrected a trivial error vehemently. No wonder that this paradoxical woman inspired love and respect—as well as irritation or resentment—and sometimes all of these, and in one person.

She demanded more of women than of men, indeed, it was her habit to cosset males. It was troubling to see her occasionally demanding approach to Elva; and sad that she could not show her granddaughters the same perceptive encouragement that had so suited the European children. It seemed wrong-headed that she sent back early letters from Elva's twin girls with grammatical and spelling errors underlined in red.

But I loved and admired Fran. I knew the kindness and delicacy of which she was capable. She was a true friend, especially generous with time and materials as I wrote this book. She had a different vision of the product than I, but I treasure the memory of the day she walked with me through the kitchen, put her hands on my shoulders, looked me in the eye and said emphatically, "Remember, Helen, it will be your book." Should I digress from my task, with talk of difficulties with one of our children or their choices, she listened with absolute attention, offered a trenchant, comprehensive comment—and we'd go back to work.

In those years after Hammy's death she often spoke of his unflappable reactions or his thoughtfulness. "Hammy set up a trust fund for me so that I wouldn't have to bother with all those accounts." Yet she maintained a firm stance of independence; she clung to life, and as certainly she found enjoyments and satisfaction. When, after a broken hip, she had to be in a nursing home in Oregon for a couple of months,

Fran in Dubai, in 1993. One of her gabboons, employed by a sheik as a falconer, hosted the visit. Here she delightedly inspects some of the falcons with their keepers.

Elva watched her become "the queen of the place. They loved her stories and her spirit. She would really have liked it there, except for the everlasting rain."

Even after the broken hip, the door to their old home remained unlocked. There were flowers on the dining room table and piles of paper on the sofa. She was cordial in greeting, busy, purposive, perhaps a bit touchier. Former boomers, early gabboons, and local friends were warmly welcomed when they stopped by. She worked hard to maintain old friendships. She kept up with correspondence, went to meetings, and provided every visitor with enthusiastic hospitality. A hunter, impressed by *Is She Coming Too?* just dropped in one day. He sat beaming at her; she invited him to lunch and ended up selling him a couple of her earlier books.

The word *loneliness* never was spoken in my presence. But that she would not admit to such feelings did not mean that she did not have them. Once, when she didn't hear my "Anyone home?" call, I went in to find her looking a bit unsettled. Perhaps she had been napping, for she said, "You didn't knock; you startled me." A neighbor woman living alone had been beaten and robbed; she may have thought a like occurrence might be in store for her. Only once did she speak of ever feeling fearful. She and Hammy had been in camp as dusk fell, and they spotted a car across the valley. Its lights showed a pattern of stopping, circling a bit, and then moving on to another searching stop. "Hammy said, 'I don't like the looks of that. Let's break camp.' We did, and got out of there fast." When she was in the hospital in 1996, not expected to live after her illness in Italy, we called her. She spoke affectionately and cheerily to me but more openly to Kip. To him she said, in a small voice, "Kip, I'm scared." And Alan Beske, who visited her in Madison at the time her eye had to be removed says that she was terrified then— and justifiably so. "But it wasn't long before she was demonstrating how to remove and insert her glass eye," he added.

She read every word I wrote. Once she exclaimed, "This is all just too sweet!" Still she wanted me to avoid certain subjects: her smoking, her tendency to exaggerate. "You'll damage my credibility!" I must not talk about her eyesight. "They'll take away my driver's license."[1] She made her displeasure clear. "You've covered the smoking already!" she said, sharply, when I quoted a friend's comment on the subject.

Fran coped with dimming eyesight and increasing deafness without complaint and made light of the occasional punies. Though she demanded much of the latest gabboons, she also gave much. She was astonishingly positive, patient, and generous. She taught them biological research methods one minute and cooked special desserts and pies for them the next. After one sudden sharp summer rain and wind, a distraught female gabboon arrived with a sodden young kestrel that had been blown from the nesting box into a drainage ditch. Fran calmly had her put it in a box, shine a naked light bulb above it, and watch it carefully until it dried. Only then should she feed it. Another girl had brought a young domestic chicken with her as a pet. Fran put up with the chaos it caused when the owl saw it and patiently helped the inexperienced girl feed and care for the gawky creature. When the girl and chicken left, she confessed: "I've never been much interested in chickens."

Sometimes everything happened at once. A lad from the nearby gladiolus farm arrived with a bushel of cut gladioli, a gabboon came back from the marsh with news of a kestrel box needing repair, the phone jangled, the pressure cooker hissed—the house seemed about to burst with demands. After one such hour, the gabboons fed, instructed, and dispatched, flowers arranged, neighbor waved good-bye with warm thanks, phone and stove calmed, Fran, then well past eighty, came smiling into the office. "Fran!" I said, shaking my head, "How do you do it?" She gave me a triumphant laugh. Interacting with young people energized her. She took pride in their demonstrations of the skills she had fostered and the affection she called forth. I once asked a gabboon how his stint up there was going. He replied, earnestly, "Nothing but good can come of this."

She was patient with them because she believed in their potential. "He will make a very good scientist, I've observed him; I know the signs," she told me gravely after I had commented on a rather childish reaction from one young man. Gabboons gave her the help and the stimulation she needed and they, in return, grew in self-confidence. One recounted her introduction to the field: "She tested whether I could orient myself to the compass directions and find nests with the aid of a map. We checked a few nest boxes but found no kestrels except for one dead bird at the bottom of a box I had reached from a ladder. I shouted the news to Fran, but against the fresh April wind she understood only 'kestrel' and decided I needed help. She plunged through the ice-cold, knee-high water in the ditch I had crossed, ladylike and dry, with my ladder as a bridge. . . . her disregard for inclement temperature is astonishing. . . . Improvisation was the rule rather than the exception. The frequent wracking of nerves and need to adapt were gladly suffered as a price for independent work."[2]

That image of the ancient Fran wading a ditch matches another: a visitor once asked if he could do anything for her.

"Yes!" she answered instantly. "You can climb up on the roof of the Volkswagen and change my yard lightbulb for me. Otherwise I'll have to do it."

"You don't do that!" he remonstrated, incredulous.

"Yes I do."[3]

She gave interviews and appeared on radio and TV. She remained the center of a circle of devoted friends and helpers. She lived with lung cancer for several years; indeed her doctors called her "the Norman

Cousins of central Wisconsin," after the noted author/editor who made popular the laughter-and-pleasure-cure for grave illnesses. She published her autobiography. She stayed in the home she so loved every summer with the help, the last two years, of Deann de la Ronde, one of her favorite gabboons and the talented illustrator of several of her books. She gave well-attended speeches to admirers within two weeks of her death. And almost everyone who came in contact with her left with the thought, "What a remarkable woman!"

But my subject wasn't Fran alone. I had read a number of her manuscripts, published and unpublished. I had seen her rehearsing her stories, and I knew she needed tell them in her own way. I wanted to bring alive a marriage to remember and an extraordinary balanced and productive team.

Those who wish a comprehensive memorial can find it in Dale Gawlik's "In Memoriam: Frances Hamerstrom" in the *Journal of Raptor Research*.[4] So in farewell to Fran—the woman who brought gifts and graces, will and emotion, to the birds, the people, the land, and the husband she so loved, I quote first from her Amazon travel journal from 1985 and then from one of her friends, who had known her longer than I had and from a different vantage point. Here she is on the Amazon and as she would like to be remembered:

Thursday p.m. [I paid] . . . roughly $7 for 1 large tough chicken. "Alfredo, do you have a knife? I want to kill and pluck the chicken."

"We do not use a knife, we do it otherwise." He paddles with one hand and gestures by spinning an imaginary chicken in the air to wring its neck.

"Well, kill it," I demand.

"There is no need." I assume the chicken is dead. Legs tied, it hangs limp. I start to pluck it. Each handful of feathers goes into the river. When I have finished with one side I turn the bird over to pluck the other side. How warm this bird remains. Plucking will be easy!

The chicken winks at me.

I gasp and look carefully. It winks again!

It is not part of my culture to cause an animal to suffer. In hasty horror I try to wring its neck. The tippy little canoe lurches wildly. I can kill a chicken so neatly with a small knife. I insert into roof of mouth, give a little twist. Chicken thrashes and dies. In desperation I try to twist its head off. Skin tears. I try harder. I consider biting it to death, but once, long, long ago I tried to bite a duck to death and discovered that a broken tooth has after effects—an enormous dentist's bill. Lightning flashes. Distant thunder. Night is falling fast. Alfredo paddles. I double the slippery neck back with all my might and break that neck apart.

Now I can see where to find more feathers to pluck only by touch or when the lightning flashes—but with each flash, sheet reflections bounce on the river and I want to see that too. Time to bail again. Here I am sitting in water, at the bottom of a "canoe" made from one tree trunk. A few drops of rain. The storm is near. It is black night on the Rio Yarapa at the headwaters of the Amazon in Peru and my clever fingers are pulling feathers out of a big, tough jungle chicken. Alfredo's paddle makes soft little splashes. We slide through the water—.

Thunder rolls along the Rio Yarapa. One can almost smell the river and the shores loom black and tall on both sides. My fingers feel for little pin feathers and if I were a different sort of person I could be in a supermarket fastidiously selecting a fryer.

And, finally, here is the testimony of a friend of over forty years. He wrote his own closest friend about visiting her as Fran recovered from her 1996 illness, and though his numbers rely on the words of a very old lady, his sentiments ring true.

Frances Hamerstrom . . . influenced me . . . deeply, as a mother for a year, when I was 14, and as a friend for life since then. . . . I took a wide, empty highway from the Central Wisconsin Airport to Wisconsin Rapids. . . . [I saw] farms with barns . . . some dilapidated. Fran told me later that farming is not what it used to be here: much of the land is in the hands of industrial food producing concerns raping the water resources for irrigation, thereby lowering the water table in the area. "And the government gives the permission!" she said.

I did not know in what shape she would be. She broke her hip in Peru last year, had it replaced, rehabilitated, went to Italy and caught a terrible pneumonia there from which she had just recovered, yet there she was, coming across the room with that smile to greet me. A little stiff, yes—but graceful, and in bare feet, as always. . . . She lived with and was married to Hammy for fifty-nine years until he died. She met him when she was eighteen, went west (literally) and worked and lived with him, always in love, until he died. She could not stand being apart from him; they did not sleep apart for more than one hundred nights during those fifty-nine years. Marriage, she said, was unimportant. They married so the children would not be embarrassed in

Fran, shortly before her death in 1998. Ursula Peterson, the photographer, went
with Fran to inspect a nearby pond soon after she took this picture. "She climbed
into my truck, we drove over and walked around the pond. She was barefoot, and
in her little dress." Both the episode and the photo show a Fran without artifice or
affectation.

school, etc. Did they "work at it?" First she denied that, accepted that "magic" was a key element. But then she revealed that there had been some principles,—for example, "never let the sun set over a disagreement." And if the sun did set, it had to be worked out in bed. "We talked and touched until it was straightened out."

We went from Wisconsin Rapids, a flat and center-less town with a paper mill, to Plainfield, passing through pine plantations. Fran: "They need a match!" (I wonder what climate change will do to those rests [*sic*] of the North American Prairie.) My heart beat faster as we turned on what used to be a dust road. It is asphalted now, cutting through forest which previously did not exist. It's the same, and it's so different. Probably it's mainly I who changed in these 41 years. The wood siding is much darker than I remember, the barn is hardly red any more . . . one of the sheds and the windmill are gone. Here I used to observe the family of woodchucks, there's where I shot at rats with a .22.

So little has changed inside! The wood stoves are all there, the water pump has been moved to another room, disorder—a symptom of creativity—prevails. This always was a household where nobody was stuck with the role of housemaid, but the hospitality was legendary; those huge breakfasts prepared for twenty people at 4:30 a.m.

After Fran had gone to bed I thought of what I had learned in this house forty-one years ago. Probably the main lesson was that it was possible not to make worries and feelings of guilt priorities in my life,—to enjoy life without automatically feeling guilty. I also learned everything about the outdoor pleasures of life: observing wildlife, taking care of animals in captivity, hunting. When I got up at daybreak Fran had been up already for some time. I went out into the gray morning to get another feel for the place, went to the open field in the back. A white-tailed deer stood there not far away. We watched each other for a little while before he moved under cover in carefully measured jumps. There was an apple tree I did not remember, it still had some apples under it in the grass. . . . On a quick tour of the roughly one square mile of Hamerstrom territory, a little piece of wilderness in the agri-industrial desert of central Wisconsin, Fran did exactly what she's always done: she let me feel what she had experienced in this place, under that tree, around that water hole: 'Here is where I saw a crane last year—the water here is much lower than it used to be—, watch it, here comes a ditch you cannot see—, those bee hives were knocked over by a bear some time ago; they belong to a neighbor who gives me as much honey as I want. I don't like honey, do you?— Yes, bears are coming southward in Wisconsin again, isn't that great? Here I used to put out carrion to attract birds of prey; the coyotes took much of it."

(Then) she gave me her address in Iquitos—just in case I pass by. Should I come to Iquitos I will look for her on the Plaza Mayor showing her glass eye to the children. When I walked back to the car she blew me a kiss and then, as I drove away, she gave me her joyful farewell signal. Raising herself to the tips of her toes, she threw up one hand and waved and smiled at me,—one beautiful woman.[5]

# 20

# A Postscript

*Is the prairie chicken hopeless?*

—Aldo Leopold, 1930

Considering this question today requires a wider point of view than when Leopold asked it. We live in a different world. Today, "sustainability" is becoming an accepted concept; national television outlets feature programs on both positive and negative environmental developments; conservation curriculums are routine in many schools; and some companies profit from green practices while others advertise their sponsorship of game or forest reserves. We hear almost daily about the release of carbon dioxide and global climate change. Respected institutes monitor the environment and inform us about such realities as the four-fold economic cost of national disasters in the 1990s over the previous four decades combined. Although we know more about the environment, we face ever more complex threats to it, among them a growing shortage of water, a continuing release of carbon dioxide, and the constant increase in population, technology, and urban sprawl. On the other hand, we see outstanding examples of individual efforts to live lives of considered conservation.

One such effort may be found at the Weaver Ranch—sixteen thousand acres of low, slightly rolling sand hills, and far, treeless horizons. It lies some fifty miles south of Portales in eastern New Mexico. The owner, Jim Weaver, met the Hamerstroms as a fifteen-year-old member of the Rockford bunch in 1958.

We were just kids. None of us liked school; home life for most of us wasn't great. Hawks saved us. Sure, we did crazy things, like our game, "hawk, owls, chase and run." Or we'd race up a tree all together, climbing over each other in our haste to reach a nest first. Yes, there were a few broken bones, but nothing really harmful. No drunkenness, no drugs. Nothing like that. I fell in love with the Hamerstrom place and the way of life. Lying on the grass out on the marsh, peaceful and still, watching the hawks—it captivated me. You heard no ugly talk. The next year, I grabbed a sleeping bag, picked up my tools, and ran away from home, hitching my way to Plainfield. After a night in a ditch with the Wisconsin mosquitoes, I was a little disheartened, but they welcomed me when I got there and I stayed a week before Hammy called me in and straightened me out. He convinced me that I should go back to school and make something of myself.[1]

Weaver did straighten out. In due time he became a gabboon, and then tried various colleges, his attendance interrupted with returns to falconry. He was hired, on the basis of experience with raptors, first by the Canadian Wildlife Service, and then by the Peregrine Fund at Cornell. Eighteen years later, in 1987, he bought a small ranch for $55 an acre near Causey, New Mexico. A remaining pocket of habitat there contained a small fraction of the once numerous population of lesser prairie chickens in Colorado, Oklahoma, Kansas, Texas, and New Mexico.

Here was the place, he thought, where he could settle and hunt the birds with his falcon. It was the start of a remarkable enterprise; one that I was able to visit in the spring of 2001. Copious spring rains had broken the drought that had stressed grasslands for the past several years. Exclamation points of golden perky sue punctuated green roadsides alongside of incandescent orange wild mallow and drifts of white tufted primroses. Suddenly, near Elida, I glimpsed the bold curves of two brown tails protruding from the thick, deep grass on the road bank. There were two lesser prairie chickens moving toward the fencerow—surely a good omen.[2]

Jim Weaver is a tall, intent, impressive man with an easy manner and a ready smile. Books have been dedicated to this mover and shaker in the falconry world, a founding member of the board of the Peregrine Fund and executive director of the Grasslands Charitable Foundation.[3] In between all this, he runs his ranch with sequential grazing and other scientific management methods.[4] Last year he bought the place across the road to save it from becoming a dairy enterprise, one that would bring in 3,600 dairy cows. Cows use a lot of water and grass. Grass, however hardy, needs water. Improper, unplanned grazing is always a danger on limited range.

Grass competes, usually unsuccessfully, on much ranchland in Roosevelt County, New Mexico, with shinnery oak, a low shrub that provides food as well as roosting and resting for chickens. Efforts with herbicides have been used to cut back the oak growth. A 1970 study found that lack of tall grass cover increased predation losses, but that few hens choose to build nests in acreage treated with herbicide. Weaver plans to use four thousand acres of his ranch in experiments that compare prairie chicken use of differentially managed plots. Varied combinations of cattle grazing and herbicide application or avoidance will provide data for analysis of this problem over a period of years.[5] He has raised significant seed money to support this cooperative project.[6]

Weaver has also imported African hawks, and cattle—Mashona, a mostly black or brown breed from Zimbabwe. Smaller than our standard breeds, they thrive in New Mexico, repaying the excruciating detail needed to get import permits and handle the red tape of several bureaucracies.

On a tour with Jim I saw outcomes of a process that began half a century ago at the Hamerstrom home. He improves habitat through sequential grazing, water management, and grassland restoration. He undertakes ongoing experiments with both wild and domestic creatures. Wildlife abounds on his acres: prairie dogs and snakes; quail, dove, and bobwhite stay year round, and the bird count was a surprising 180 species last year.

We stopped at what looked like a small junkyard—abandoned metal in a bare patch of ground. "Look at those circular structures with the round holes." I saw the unblinking eye in the circular face of a barn owl on her nest in an abandoned "heater-treater," as locals dub one of the devices used in oil field installations. "The lessee is supposed to pay all costs of cleaning this up, and 'restoring' the land to how it was before they got here," Jim shrugged. "We did, at least, get them to clean up the pools of residue. Otherwise we'd find up to seven young owls, dead, in the pools that they mistook for water."

Two fences and two gates later we came to a shallow lake. Jim maintains its water level with a small constant trickle of water from his pipe-and-pump system. "This is part of the La Playa lakes, a system of former wetlands, now almost all dry, that stretch well into Texas. We've found Indian artifacts," Jim said, matter-of-factly. "They probably had seasonal camps here." Cattails rose from the gleaming expanse, an avocet waded, mallards fed. As we breathed in the peace of this oasis, we saw a truck roll slowly down the ranch road toward us. A herd of cattle walked calmly behind the truck: it had a feed bin mounted at the rear. "We're moving them to new pasture," Jim said. "They'll be there for a couple of days; then we'll move them again." Nearby we saw a two hundred-acre prairie dog town, one of six flourishing on his spread. Miles of fence border selected individual pastures, miles of plastic pipe provide water to each one, for stock and wildlife needs and some irrigation.

Back at his home, dozens of boat-tailed grackles chuckled in a huge Chinese elm, rising and diving over the green expanse at his back door. White-necked ravens rode the wind. Weaver's young son was fly-fishing in a stocked pond close by. "Two big bass in there must have been caught forty times. We hook and release." An array of large metal sheds and barns rose near the homestead. As we headed toward his falcon barns, we passed large mounds of sacks. "Native grass seed; the stuff's expensive," he commented. "It needs rain to germinate—so we're seeding now. That's what that airplane was doing that you saw across the road."

We reached a closed door. "Wait till your eyes adjust inside, then you can see the birds through those little one-way-glass windows. Speak very quietly. We have four breeding pairs." We saw the hawks, and then his incubator/brooders, where ingeniously engineered, climate-controlled Plexiglas contraptions hold eggs, and, as a small sliding panel revealed, a downy teifa falcon chick. It gaped as the panel moved.

Before settlement, tall and mixed grass covered the entire Llana Estacado—a high plateau along the Texas–New Mexico border. Conrad Richter, whose novel *The Sea of Grass* describes the conflict and tragedy, both personal and environmental, between eager settlers (he called them "nesters") and a gentleman land baron, wrote of a "shaggy" prairie. Richter's young narrator describes it: "[I saw] the

grass in summer sweeping my stirruped thighs and prairie chicken
sculling ahead of my pony; with the ponds in fall black and noisy with
waterfowl, and my uncle's seventy thousand head of cattle rolling in
fat; with the tracks of endless game in the winter snow and thousands
of tons of wild hay cured and stored on the stem; and when the
sloughs of the home range greened up in the spring, with the scent of
warming wet earth and swag after swag catching the emerald fire . . .
and everywhere the friendly indescribable solitude of that lost sea of
grass."[7]

Settlers broke the sod on their 160-acre allotments. At the first
cycle of drought, they went broke trying to grow corn and wheat as
they had back east. As the flood of nesters gave up, the cattlemen who
had survived their invasion took over the damaged range sometimes
overgrazing and further degrading its wind-eroded surface.

Richter's novel unwittingly prefigures an ideal of progressive modern ranchers: healthy grass, abundant wildlife, fat cattle, and—of
interest to friends of the grouse species—countless prairie chickens.
Thus he works toward reaching Leopold's standard: "When the land
does well for its owner, and the owner does well by his land; when both
end up better by reason of their partnership, we have conservation.
When one or the other grows poorer, we do not."[8]

As we drove over swells of shinnery oak, I heard details of Jim's
part in the saga of Nancy, Fran's eagle. He, his brother, and Rod Friday found the eagle in 1963 in Wyoming, as a wretched chick—small,
tick infested, unlikely to survive in the nest in competition with a
stronger sibling. They saved her; Jim kept her at his father's Illinois
farm for nearly a year.

The Feds showed up. Someone had talked, I suppose. They told me they must
take her. I refused until they convinced me that she would be well cared for in
Lincoln Park Zoo, where she'd be quarantined briefly, and then placed in good
facilities. They wanted me to give her to them; I wouldn't let them have her.
Instead, I took her there myself and turned her in.

But when I went back to visit her, I found her on cement, in dark quaran-
tine quarters, six feet square. There was a little window, high up; she could
never see the sun. She was in heavy molt. I called Fran. My Aunt was a friend
of the Director's, and within twenty-four hours Nancy was released to Fran's
care, for rehabilitation.[9]

Weaver's account demonstrates character—perseverance and clear purpose, force of personality, and practical know how in meeting obstacles—traits that serve a conservationist well. I realized, there at the ranch, that his situation is more like Hammy and Fran's in Plainfield than at first appears. Weaver, like them, is an advocate armed with useful but incomplete understanding of the interaction of complex factors. He, like them, tries to save a declining species threatened by agricultural development, technology, and habitat degradation in a place of limited rain and surface water. He has turned, as they did, to doing and using transformative research and building support for it. He, as they did, multiplies his dedicated and costly personal involvement by collaborating with national, state, and local groups. He, like the Hamerstroms, uses hospitality to educate: "I never expected to have three hundred and fifty visitors in my home, but that's what came through here last year," he reported. He, too, understands the need for political action: he recently helped prevent the listing of the lesser prairie chicken on the endangered species list. "How can it help when there's no money, if they are put on the list, to do anything about the real problem, which is habitat loss!" Above all, he looks to the future; he has bought most of the small, moribund town of Milnesand with an eye to restoring a large house to serve as a headquarters for researchers and observers at the state prairie chicken areas and nearby ranches. In 1992 he was appointed to the New Mexico Game Commission.

Aldo Leopold believed that the farmer must be the cornerstone of conservation efforts. Not many years later, as agricultural patterns changed, the Hamerstroms had to resort to state purchase of scattered land for management instead of working with individual farmers. Now, as small farms continue to give way to giant corporate enterprises, Weaver returns to the idea of individual action. Indeed, he believes, as do an increasing number of others, that ranchers are uniquely situated to be the foundation of an effective conservation movement. The livelihood of a rancher depends on productive land, and the very processes that sustain habitat create healthy watersheds, increase diversity, and build good ranches. Weaver was pleased with a survey that scored ranchers and environmentalists close together at the high level on a scale of conservation beliefs, whereas farmers as a class, not as individuals, place well below them.

The Weaver Ranch gave me a new outlook on Leopold's query. His 1930 question must now apply to the entire United States grouse species, including the lesser prairie chicken, the Attwater's prairie chicken, and the sage grouse.[10] And although the picture painted in *The Greater Prairie Chicken: A National Look* may seem to indicate serious doubt for the species' survival, I see it as a call to practical action.[11]

One hundred years ago the greater prairie chicken inhabited eighteen states and four Canadian provinces. Today it is found in only eleven states and not at all in Canada. No simple summary of the status of each state provides clarity; too many factors beyond human control intrude. A look, however, at the extremes gets one's attention. In Texas, the Attwater's prairie chicken, found in the thousands between the Nueces River in Texas and Abbeyville, Louisiana, in the past, was declared endangered after a 1967 count of a slim 1,070 chickens in fourteen Texas counties. Vigorous conservation efforts and wide-ranging research may have contributed to a rise to 1,600 birds in 1980,[12] but, according to Dr. Nova Silvy, the principal Attwater researcher, brush encroachment, urban and industrial growth, and drought took their toll and habitat decreased through the 1980s. No truly wild stock now remains in Texas. Other states may learn from this cautionary case history, for twenty-five years ago the situation in that state did not seem dismal, as it now does.[13]

In contrast, Colorado—where the hunting season closed in 1937, and a population decline continued from about two thousand greater prairie chickens in the 1950s to between seven and eight hundred in 1964 and to the designation "endangered" in 1973—demonstrates success. After the Colorado Division of Wildlife began protective conservation policies and the translocation of birds, a steady recovery has created habitat for between eight and ten thousand of the birds, now classified as nongame wildlife.

Hunting seasons have been severely restricted or closed for this sought-after game bird in several states (such as Oklahoma) where hunting was routine in Hamerstrom days. The species is listed as endangered or threatened in four states, protected in four others. No simple, time-bound statement of status gives a real picture of the potential and pitfalls inherent in chicken population estimates. States with apparently stable prairie chicken populations such as Wisconsin and Minnesota illustrate the importance of large state-managed acreages and Conservation Reserve Programs grasslands. These states are sites for research about a reliable guide to minimum viable populations, and indeed, the flow of research does not abate.[14] Interconnected ranges may provide hope for states such as Iowa and Missouri, where trends have been downward. On the other hand, isolated Illinois, in spite of vigorous efforts at translocation to supplement a small population, may fail to establish self-perpetuating flocks. Isolation from other populations is a potential genetic bottleneck. And in Nebraska, where a decline was halted by an expanded Conservation Reserve Program that returned some pivot-irrigation acreage into fallow land, a return to overstocked, overgrazed land could again threaten the once chicken-rich sand hills. Private organizations such as the Nature Conservancy are making an impact in Oklahoma; South Dakota shows the beneficent effect of Conservation Reserve Program acreage—a change that may vanish with the next whim of Congress or significant change in the agriculture demand/supply situation.

In sum, it all depends on land-use patterns and the tenacity, skill, and will of citizens and sportsmen, and of private, state, and national agencies to balance the negative trends in farming and cattle raising; the complexity of needed research, the trends in population, and the continuing effects of technology and industrialization indicate that the future of the North American grouse species will be difficult. This statement was recently confirmed by an article in the *National Geographic* of March 2002. The history and problems facing the various subspecies are explained and clarified by cogent discussion along with splendidly designed and presented maps. More information can be found on the magazine's website.[15]

Difficult it may indeed be, but no more difficult than the task of restoring degraded western ranchland, reintroducing wolves, or halting the shocking shrinkage of chimpanzee numbers in Africa. A cautious optimism seems justified by the growth in environmental awareness among both the public and policy makers. Consider the growth of conservation-minded groups that publish bulletins and newsletters, and the burgeoning local organizations of all kinds that try to deal with problems at many levels.

When the Hamerstroms started work on the Buena Vista marsh in the 1950s, no *New Ranch Handbook* existed.[16] They had no manual that shared models of successful methods of habitat improvement with groups at odds with one another. The Hamerstroms could not turn to experts in conflict resolution or read the academic studies on collaborative approaches available today.[17] They had, in the face of considerable difficulty, to lay down the fundamentals that made possible the varied approaches to today's grouse management. They had to build a foundation for patterns of research and its applications for managing grassland species. An overview of *The Journal of Raptor Research*'s commemorative issue of September 1992 makes clear the details and significance of the Hamerstroms' life and methods. It seems safe to assert that their encouraging guidance and support and their vivid example nurtured many into science and art. Now people who knew them, like Joseph Schmutz, can assess "Hamerstrom Science," and its importance to public understanding; ex-gabboon Dale Gawlik can justifiably identify Errington and Hammy as pioneers in the enlarging field of conservation biology.[18]

Today, in contrast, innovative approaches to saving habitat abound. The MIT Press is publishing books describing diverse methods of conservation: new materials for building, reduction of the trash burden, plant breeding to reduce pesticide and dye use, and even rooftop agriculture that could completely remake food production and delivery in cities.[19] Collaboration is common. The recent conference on environmental conflict resolution at the Udall Center for Studies in Public Policy at the University of Arizona drew a healthy 178 attendees.

The Hamerstrom's dedication, their model of demonstrating that personality, principle, and solid science can fertilize, as it were, the growth of previously unheard of possibilities. Like the sower in the biblical parable, their seed—and that spread by those they touched, along with others influenced by the many and varied forerunners of today's conservation movements—has fallen upon fertile ground and multiplied a hundredfold. The burgeoning of those seeds may, in the fullness of time, bring about a restoration of purpose and reconciliation between human requirements and ecological well-being.

# Notes

In citing works in the notes, short titles have generally been used. Works frequently cited are identified by the following abbreviations: FH, Frances Hamerstrom; FNH, Frederick Nathan Hamerstrom Jr.; AL, Aldo Leopold; CM, Curt Meine.

I have preserved the original punctuation, date as given, and syntax of the original letters.

### Chapter 1. Prologue

1. Modest Hammy would have nothing to do with a mixed sauna.

2. Following Fran's death in 1998, their daughter, Elva Paulson, moved books, family papers, and other material to her home in Oregon; what professional files remain are still in the Plainfield house. There is no organized archive at the present time.

3. The approach of Italian scholars Alessandro Portelli and Luisa Passerini, among others, has brought new impetus to the use of oral history. Denying its reliability is no longer automatic: reliability is not everything. "Oral sources tell us not just what people did, but what they wanted to do, what they believed they were doing, and what they now think they did." See Alexander Stille, "Prospecting for Truth in the Ore of Memory," *New York Times,* 10 March 2001, A15, A17.

4. April 9, 2002 Fran wrote on an early draft of this work: "This rings true so often—Don't be afraid of lots of direct quotes—Paraphrasing weakens the impact."

5. Aldo Leopold, *Game Management* (New York: Charles Scribner's Sons, 1933), 423.

6. Ruth Hine, personal communication.

### Chapter 2. The Complexities of Childhood

1. Frances Hamerstrom, *My Double Life* (Madison: University of Wisconsin Press, 1994). The Press reprinted her earlier *Strictly for the Chickens* (Ames: Iowa State University Press, 1980). Citations are from the more recent book.

2. See Edward F. Flint Jr. and Gwendolyn S. Flint, *Flint Family History of the Adventuresome Seven*, 2 vols. (Baltimore: privately printed, 1984), 2: 1029–1030, 1048–1049; and Erwin James Otis and Florence Leverett Hodge, *Genealogy of William Leverett (1772–1807) and Descendants* (Ann Arbor: privately printed, 1974), leaves 4 and 5. I am indebted to John Holshueter for providing these references.

3. Dates and facts derive from two detailed letters to me from Putnam Flint, 22 January 1999 and 19 August 2000.

4. This was the Herr Thümmer who came to a Sunday tea with the young Flint family in Dresden, probably in 1912, a visit that lay behind Fran's discovery that her father had "condoned breaking the law." See FH, *My Double Life,* 5.

5. Ibid., 81. Putnam Flint was distressed when Fran sold this house, a bequest from her grandmother, for $10,000 shortly after World War II. It was valuable ocean-side property.

6. Putnam Flint has this diary.

7. FH, *My Double Life,* 12.

8. Alice Miller, *For Your Own Good: Hidden Cruelty in Child-Rearing and the Roots of Violence* (New York: Farrar, Straus, 1986), 97.

9. FH, *My Double Life,* 26; Mrs. Flint's letter to Fran is in Putnam Flint's family papers.

10. Ibid., 40. "I sadly repent all my nauty-nesses to Fräuta but perhaps it is too late she talks every day about going and I am very unhappy. I have resolved to be good." The method of adult control implied here is laying on guilt, a common practice no longer seen as desirable.

11. Both Fran and Helen Flint kept in touch with Fräuta until 1941, and again after the World War II.

12. Hamerstrom's accounts varied. In her *Birding With a Purpose* (Ames: Iowa State University Press, 1984), 4, she says: "My family graciously gave me a vacant maid's room for my hobbies. It contained my insect collection, my animal collection, my bird collection, my egg collection, arsenical soap for preserving skins, and things that I just happened to like: for example, a doll's bureau with a secret compartment for hiding small objects." This account serenely contradicts her frequent statements about the secrecy of her pursuits.

13. FH, *My Double Life,* 116–117.

14. Ibid., 71–72. See "Will She Die without Rangle?"

15. Her wartime letter to Elva of 16 March 1944 to McCool Junction, Nebraska, reads, in part, "Beloved little Elva—Happy birthday darling!" She mentions the check she is sending to Fran for a birthday present and continues, "Grandparents have so much love for you and Alan that sometimes they find it hard to be brave in Mother's old home where the beautiful snow is all around and it is impossible for the precious Hamerstroms to come. Devoted love, Treasure, Grandmother."

16. FH, *My Double Life*, 97–100.

17. Putnam Flint to FH, 1 March 1990.

18. FH to Putnam Flint, 12 June 1983.

19. Our family visited him and his wife at his creatively remodeled farmhouse in the Catskills the next summer. Characteristically, Fran stormed, "They ruined that house!" Davis, a highly regarded architect, died in 1993.

20. Elva Hamerstrom Paulson to author, "I never heard that story from my Father."

21. Ruby Darrow to her sister-in-law, Helen Darrow, dated "Later, Feb-21."

22. The Hamerstroms reportedly went down to move this uncle into a retirement home in Galesburg, Illinois. Fran occasionally used his letterhead.

23. Irving Stone, *Clarence Darrow for the Defense* (Garden City, N.Y.: Doubleday Doran, 1941), 162.

24. Burt Hamerstrom to "Dear Fred," 6 June 1904.

25. *Germantown Telegraph,* undated clipping. Edward Philips, a local historian and columnist, used Louise Davis's letter of January 1927 to augment his own observations on progress.

26. Even the adult Frederick Hamerstrom was erratic about spelling, contractions, and apostrophes.

27. Her account implies that Fran and Hammy returned to his parents' house, presumably after Hammy's mother's 1939 death. Putnam's letter of 19 August 2000 said that Fred Sr. lived for some time with another family in Winchester before he died in 1948.

28. Davis Hamerstrom to author. The tape he provided makes almost tangible the sensations of their boyhood in the then semirural environment of their homes.

29. Ruby Darrow to Helen Darrow, 21 February 1940.

30. FNH to his parents, undated.

31. Fran put it into students' heads to "Think lemonade" whenever they heard the word *water* in the daily Bible reading. (*My Double Life*, 89). As an alumnus of a missionary school in India, I remember our jejune delight at adding the phrase "under the sheets" as the first lines of the frequent hymns were announced.

### Chapter 3. Self-Discovery and Love

1. Much of the information in this chapter derives from Thad Smith's conversations and phone calls to me.

2. A card in Fran's address file from the office of admissions recorded this information.

3. He told Fran that he quit this job because he didn't like being tipped by prostitutes.

4. Harvard Placement Service form, 1931.

5. FNH to his parents, undated.

6. FNH to his parents, 13 November 1927.

7. Davis concurred, acknowledging Darrow's help. "I wouldn't have been able to go to M. I. T. without it."

8. Kevin Tierney, in *Darrow: A Biography* (New York: Crowell, 1979), dates this marriage in 1903, well before Frederick Hamerstrom's birth.

9. Ruby Darrow to her sister-in-law, Helen Darrow, undated.

10. FNH to his parents, undated.

11. FNH, application to Harvard, summer 1928.

12. Ruby Darrow to Helen Darrow, undated.

13. FNH to Henry Pennypacker, 30 August 1929.

14. Alan Hamerstrom, their son, has a different take on the matter. "Oh, Miss Swanson may have said something like, 'You should try out for the movies.' Fran's memory probably did the rest."

15. FNH to his father, undated.

16. FNH to his parents, from Iowa after he passed his master's exam there in 1936.

17. Frances Hamerstrom, *Is She Coming Too? Memoirs of a Lady Hunter* (Ames: Iowa State University Press, 1989), 26.

18. Elva Hamerstrom Paulson smiled as she told this story. She had reason to smile. The distress that her parents felt at her choice of a husband was also clearly, though more gently and briefly, expressed. See chapter 14.

19. FNH to his parents, undated.

20. Fran, at eighty-five, supplied this name without missing a beat. The signature, A. Harrison Loache, appears on their first (Florida) marriage license.

21. Fran's unusual storage spot for this document was in the almost blank baby book that recorded her children's birth dates.

22. Frederick N. Hamerstrom Jr., "Top Honcho," in Richard E. McCabe, ed., *Aldo Leopold, Mentor* (Madison: University of Wisconsin Department of Wildlife Ecology, Aldo Leopold Centennial Symposium, 1988), 32.

23. FH, *Is She Coming Too?* 33–34.

24. The return address was c/o Bankers Trust Co., 5 Place Vendome [Paris], 28 March 1928.

25. Harvard Placement Service, "Record of Employment," 26 February 1931.

26. Another relative summed it up: "Why would he want to go there after the way they treated him!"

### Chapter 4. Students, Teachers, and New Horizons

1. They taught whatever parents requested. Fran wrote of learning more algebra in two days than she knew for college boards, "before starting to tutor Pat. He is an awfully nice chap." Pat invited her to crew for him in sailing races. FH to her mother, 2 August 1930.

2. Richard Wentz, ed., *Return to Big Grass* (Long Grove, Illinois: Ducks Unlimited, 1986), 3.

3. FH, *Is She Coming Too?* 42. Would rats inhabit the home of wealthy people in Milton, Massachusetts? Fran often said so.

4. Ibid., 51.

5. Beauveau Beals, personal communication.

6. FH, *Is She Coming Too?* 60.

7. McCabe, *Aldo Leopold, Mentor,* 33.

8. Curt Meine, *Aldo Leopold: His Life and Work* (Madison: University of Wisconsin Press, 1988), 277.

9. Paul L. Errington, *The Red Gods Call* (Ames: Iowa State University Press, 1973), 37. This information on Errington derives from this book and its introduction, vii–ix.

10. Pauline Drobney, "The Phoenix People of Sod Corn Country." In *Recovering the Prairie,* Robert F. Sayre, ed. (Madison: University of Wisconsin Press, 1999) 167.

11. FNH to his parents, undated. His letter of application went to Errington on 2 September 1932.

12. FNH to his parents, "Friday."

13. Harold Peterson, "Prairie Chicanery," *Sports Illustrated* (24 May 1967): 47–52.

14. FH to FNH's parents, undated.

15. Judy A. Minnick kindly provided transcripts from Iowa State College to Elva Paulson on 10 October 2000. Hammy got his M.S. degree 10 June 1935; Fran, a zoology major, graduated with a B.S. in March of that year. Her letter to her mother from Ruthven, their summer research station lists some of the courses she especially liked.

16. Actually, she became a rather good amateur painter. Her work appeared on a cover of the *Passenger Pigeon,* and her paintings regularly won first place at the Waushara County Fair, until the organizers asked her to stop submitting so that other local artists would have a chance to win.

17. Carolyn Errington to author, 8 August 1992. Ada Hayden was primarily responsible for the first purchase of land for today's preserve.

18. Ibid.

19. Paul L. Errington, Frances Hamerstrom, and Frederick N. Hamerstrom Jr, "The Great Horned Owl and Its Prey in North-Central United States," Res. Bull. 277, Iowa State College of Agriculture and Mechanical Arts, (1940): 758–850. The paper lists over seventy-five species of vertebrate and invertebrate remains identified in the pellets. It won the Wildlife Society's award for that year.

20. I saw her sharing a table covered with insect remains with a young person whose face showed considerable distaste for the job.

21. Frances Hamerstrom, *Harrier, Hawk of the Marshes: The Hawk That Is Ruled by a Mouse* (Washington: Smithsonian Institution Press, 1986), 26.

22. My colleague at the University of Wisconsin–Stevens Point, Jack Heaton, repeated a story told him by wild turkey researcher Bob Jonas, later a wildlife biologist at Washington State University. In the late 1960s, the vacationing Hamerstroms went into Ekalaca for a Saturday night beer. They bought rounds for all, cowboys and ranchers included, and around midnight, asked where they could camp. Concerned that the pair had used up all their cash in their largesse, the group took up a collection to pay for a hotel. These hardy souls couldn't believe that these two eminent people *preferred* to camp.

23. FNH to Clarence Darrow, from Ruthven, Iowa, 3 August 1934.

24. FNH to his parents, undated.

25. FNH to his mother, undated.

26. FNH to his parents, June 1936.

27. Frederick N. Hamerstrom Jr, "A Study of the Nesting Habits of the Ring-Necked Pheasant in Northwest Iowa," *Iowa State College Journal of Science* 10 (1936): 173–203.

28. FNH to his parents, undated.

29. FNH to his father, undated.

30. Ruby Darrow to Helen Darrow, 4 March 1934 or 1935.

31. Putnam Flint, personal communication.

### Chapter 5. Conservation Beginnings in a Midwestern Appalachia

1. The village population in 1930 was 761, the rural farm population was 520. It was classified as "foreign or mixed parentage," "foreign born, white," and "native white." *Fifteenth Census of the United States: 1930. Population,* vol. 3, pt. 2 (Washington D.C.: Government Printing Office, 1932), 1354.

2. "History of the Farm Security Administration" (U.S. Department of Agriculture, Farm Security Administration, 10 October 1940, mimeographed), 12 pages. Readers interested in the Great Depression may wish to

know that copies of the Resettlement Administration reports and correspondence are probably stored in the Midwest Archives in Chicago, of the National Archives and Records Center. Record Groups 96.4.1 through 96.4.12 might well be worth exploring. I thank Jack Holzhueter, formerly of the *Wisconsin Magazine of History,* for this information, which he sent me in 1999.

3. James Agee and Walker Evans, *Let Us Now Praise Famous Men* (Cambridge, Massachusetts: 1960), 35.

4. Rudolph A. Christensen and Sydney D. Staniforth. "The Wisconsin Resettlement Program of the 1930s, Land Acquisition, Family Relocation, and Rehabilitation," *University of Wisconsin Department of Agricultural Economics* 52 (May 1968).

5. Ted H. Watkins, *Righteous Pilgrim: The Life and Times of Harold L. Ickes, 1874–1952* (New York: Henry Holt, 1990). Pages 250–53 vividly detail the conditions calling for redress.

6. "History of the Farm Security Administration."

7. Michael Goc, "The Wisconsin Dust Bowl," *Wisconsin Magazine of History* 73 (Spring 1990): 169–171.

8. AL to Taft, 5 March 1934, as per his copy to Hammy along with a copy of his prompt reply to Taft of 10 March 1935 from the New Soils Building on the Madison campus: "In cooperation with . . . the Conservation Department, we already have a couple of men making a survey of the marshlands in your part of Wisconsin with a view to recommending them as a federal project." Below his signature is the designation, "In charge, Game Research." Leopold Archives, University of Wisconsin Archives, ser. 9/25/10–1, box 3.

9. "The Wisconsin Dust Bowl" makes clear the dismal history of the drainage districts that covered—and degraded—some four hundred thousand acres of central Wisconsin.

10. Ibid., 175. Goc's source, Marilyn Cofey in *Natural History* (February 1978) says that storm led to Roosevelt's promise to plant six thousand miles of trees in the prairie states.

11. Frederick N. Hamerstrom Jr., "Management Plan for the Central Wisconsin Game Project: (LD-W-5), 6–8.

12. Goc, "The Wisconsin Dust Bowl," 167.

13. "Wisconsin Resettlement," 46–47. Another short mimeographed report in Hammy's files, L. G. Sorden's 1938, "Northern Wisconsin Settler Relocation Project," cites the $15,000 per year saved by closing schools and cutting relief costs. He praised the "renewed hope given to these people."

14. Fred Zimmerman, personal communication.

15. Goc, "The Wisconsin Dust Bowl," 163, quoted an article, "If We Get Rain," by famed journalist Paul de Kruif in the *Ladies Home Journal,* September 1934.

16. His extended family eventually owned some one thousand acres of irrigated land on the Leola marsh.

17. Studs Terkel, *Hard Times: An Oral History of the Great Depression* (New York: Random House, 1970), 262.

18. One of my most vivid memories is of the answer an illiterate mechanic—who could build anything needed at the University of Wisconsin Primate Lab—gave me when I asked him, "Who were good presidents?" He scowled, and declared, "There has only been one President! You know who—Roosevelt." Another is of a brown-eyed eleven-year-old named Nira at a Girl Scout camp near Flint, Michigan, in 1945. I asked the source of her name. "The National Industrial Recovery Act," she replied proudly. "It gave my Dad a job."

19. FH, *My Double Life,* 121. Cox, as the first forester in Minnesota, had created that state's Forest Service. See *Minneapolis Star Tribune,* 26 December 1999, 10.

20. FNH to Franklin S. Henika, 18 October 1935.

21. Frederick N. Hamerstrom Jr., "Wildlife Research on the Central Wisconsin Game Project." He asked that this eleven-page report be read at a meeting of the Resettlement Administration Conservation Unit in Milwaukee, December 1936.

22. FNH to W. T. Cox, Regional Forester-Biologist, 31 December 1936, 3–4.

23. FNH, "Wildlife Research . . . Game Project," 2: The "Report of the Fur Inventory" (LD-WI-5, 23 August 1937), records numbers of furbearers seen from 6 April through 28 July, in 1,800 man-hours at a cost of $862.26.

24. Ibid.

25. In the late 1950s I was told that the best teacher in the Hancock grade school was—gasp!—"living in sin." I found that, forced to choose between a legal marriage to a struggling farmer and a job, she had sensibly simply moved in with him. Jobs were for men; married women were fired in that school district in the 1930s.

26. FNH to his parents, fall 1936.

27. FH, *My Double Life,* 125.

28. Ibid., 127.

29. FNH to R. I. Nowell, 21 January 1937.

30. They established the bearings of the landmarks, located them exactly, and supervised crews to ensure the accuracy and hence usefulness of the finished product.

31. Art Hawkins kindly looked up his "Game Research News Letter" file of 1936 and 1937 and made this information available to me.

32. AL to Silas J. Knudsen, 20 July 1937.

33. FNH to his parents, spring or summer 1937.

34. U.S. Department of the Interior, *Necedah National Wildlife Refuge*

*50th Anniversary* (Washington, D.C. U.S. Fish and Wildlife Service, 1989), 1, 34, 35, 38. A faint record of Hamerstrom toil remains in the list (40) of species planted in 1936 and 1937.

35. U.S. Department of the Interior, *National Wildlife Refuges: A Visitor's Guide* (Washington, D.C.: U.S. Fish and Wildlife Service, n.d.).

**Chapter 6. Enter Leopold and the Chickens**

1. CM, *Aldo Leopold*, 304. Leopold's *A Sand County Almanac* contains many expressions of this philosophy in the original introduction, in "Thinking Like a Mountain," and in section 4 of the expanded edition, "The Land Ethic."

2. Watkins, *Righteous Pilgrim*, 491.

3. CM, *Aldo Leopold*, 306; Aldo Leopold, *A Sand County Almanac* (New York: Sierra Club/Ballantine, 1974), 138.

4. CM, *Aldo Leopold*, 310.

5. McCabe, ed., *Aldo Leopold, Mentor*, 51.

6. Roger Tory Petersen, an eminence in the bird world, would write the foreword to *Harrier, Hawk of the Marshes* years later.

7. CM, *Aldo Leopold*, 378.

8. McCabe, *Aldo Leopold, Mentor*, 40.

9. CM, *Aldo Leopold*, 391.

10. McCabe, *Aldo Leopold, Mentor*, 21. Curt Meine reports that over thirty of his short pieces appeared in *The Wisconsin Agriculturalist and Farmer* in a four-year period.

11. This dictum became Fran's mantra.

12. FH, *My Double Life*, 141–144.

13. McCabe, *Aldo Leopold, Mentor*, 63.

14. Ibid., 34–35.

15. AL to prospective graduate students, 2 February 1942. He cited two of Hammy's papers published in 1939 and 1941 in *The Wilson Bulletin*. Leopold Archives, ser. 12/20/45, box 001.

16. Frederick N. Hamerstrom Jr. and James Blake, "Winter Movements and Winter Foods of White-tailed Deer in Central Wisconsin," *Journal of Mammalogy* 19, no. 2 (1939): 205–15.

17. Fran was proud of his foresight and often remarked on the wisdom of his prediction.

18. Leopold had strong feelings about the training and certifying of wildlife professionals. As he wrote on 15 October 1938 to Rudolf Bennitt of the University of Missouri: "I opposed the attempt to build a fence around our membership in the first place, and I am afraid I do not regard as vital the many questions now arising as to the maintenance of that fence." He was serving on what must have been a difficult assignment to a committee of the

Wildlife Society on "Standards of Training" for the new profession. That he considered Frederick and Frances Hamerstrom to be among the "four best" of his students, as Leopold wrote Bennitt on 16 January 1939, again indicates his belief in the value of practical experience in the field. Leopold Archives, ser. 9/25/10–2, box 9.

19. Joseph Hickey, personal communication.

20. An active range specialist recently deplored the difficulty of working with graduate students, finely trained in statistics, computer modeling, and laboratory techniques, yet who, in the field, express fear that "a prairie chicken will bite them."

21. CM, *Aldo Leopold,* 360. He was quoting Karl T. Frederick, in a 1935 issue of the *Rifleman.*

22. An undated letter to his parents summarizes areas he thought would concern them: "This month, I am trying to finish writing up the Resettlement Administration for publication (6 or 7 papers). . . . I have a University fellowship for next year—$600—which will help greatly. We had saved enough money for two years in school, but found it would take three. . . . The fellowship, by the way, is a reward of virtue, in this case for good marks. . . . My graduate record for three years at Iowa and one semester here, you may be interested to know, is 2.8. 3.0 is all A's."

23. CM, *Aldo Leopold,* 348.

24. Schmidt's career began when A. O. Gross, an authority on the heath hen and who was hired in 1929 by the Wisconsin Conservation Department to study the Wisconsin prairie chicken, hired him as his assistant.

25. Leopold Archives, ser. 9/25/10–5, box 8.

26. See a drawing of this trap in *My Double Life,* 205.

27. FH to Gustav Kramer, 20 February 1948.

28. FNH to his parents, undoubtedly in the fall of 1938.

29. FH and FNH to his parents, December 1938.

30. Grange, who was Wisconsin's first Superintendent of Game, was cousin to Red Grange, the famous "galloping ghost" quarterback of Notre Dame. He published an acclaimed book about the life of animals, *Those of the Forest,* in 1953. Fran very much admired it. "Was Grange prickly?" I asked her. "Prickly but lovable," she replied.

31. A. W. Schorger, "The Prairie Chicken and Sharp-Tailed Grouse in Early Wisconsin," *Transcript of Wisconsin Academy of Science* 35 (February 1947): 1–57.

32. Wallace Grange, *Wisconsin Grouse Problems* (Madison: Wisconsin Conservation Department, 1948), 19–29.

33. FH, *My Double Life,* 154.

34. Ruby Darrow to Helen Darrow, 21 February 1940.

35. "Pine Plantation Is Pioneer Memorial," *Oshkosh Daily Northwestern,* 3 March 1976.

36. F. B. Trenk, "Oldest Plantation Yields Sawlogs." *Wisconsin Conservation Bulletin* (Jan.–Feb. 1962). Mr. Jerry Carleton of Hancock kindly provided me with a copy of the piece.

37. Ibid.

38. FNH and FH, "The Last Year of the Pigeons Was Terrible," *Passenger Pigeon* 37, no. 2 (Summer 1975): 55–56. The authors were cited as F. and F. Hamerstrom.

39. Justin Isherwood, "Who Rules the Marsh," *Stevens Point Daily Journal,* 28 July 1977.

40. George Rogers, *Stevens Point Journal,* 20 August 1982.

41. Goc, "The Wisconsin Dust Bowl," 169.

42. Things were no different in 1980. Our rural mail carrier, determined to farm, rented land from the Department of Natural Resources, now managing parts of the marsh. A few years later, he gave up. "You just can't make it. One good year in four or five doesn't make up for those frosts 'most every year. Why, I couldn't even salvage silage!"

43. FH, unpublished manuscript.

44. A cleared eighty acres that we bought in 1959 at $47.50 an acre had sold for $7.62 an acre in 1928. The corner of that piece of land had been leased to Joint School District #3 for $6 per year, "as long as used for school or public purposes." In 1937 an adjoining eighty acres sold for $300; and in 1941, on the frequent delinquent tax sale, a price of $21.10 an acre was recorded.

45. Leopold called hunting chickens the "grand opera" of sport. His reluctance to kill predators began as he watched the green fire in the eyes of a dying female wolf. *A Sand County Almanac* records the experience and his comment: "I was young then; and full of trigger-itch."

46. Nina Leopold Bradley, personal communication.

47. Author and Fran Hamerstrom, personal conversation.

48. FH, *My Double Life,* 172.

49. FH, *Birding With a Purpose,* 6–8.

50. Frances Hamerstrom, "Dominance in Winter Flocks of Chickadees," *Wilson Bulletin* 54 no. 1 (1942): 32–42.

51. FH, *My Double Life,* 188.

52. Joseph Hickey, personal communication.

53. FH, *Birding With a Purpose,* 35–36.

54. Frederick N. Hamerstrom Jr., "A Study of Wisconsin Prairie Grouse: Breeding Habits, Winter Foods, Endoparasites, and Movements" (Ph.D. diss.; University of Wisconsin, 1941).

55. Hazel Grange, *Live Arrival Guaranteed* (Boulder Junction, Wisc.: Lost River Press, 1996), 256. They were facing foreclosures on land and machinery when Wallace's first place in the civil service exam led to his being hired. "His salary, shortly, was spread far and thin."

56. FNH to Donald Douglass, Michigan Conservation Department,

19 November 1942. The copy that Hammy sent to Leopold contains his handwritten notes.

### Chapter 7. An Interruption

1. FNH to his mother from Ames, therefore before 1937.
2. Putnam Flint's wife, Dorothy Ann, is certain of this information, repeated in several phone conversations on the subject.
3. FH, *My Double Life,* 139–140.
4. As Leopold himself reported in his second "Progress Report," of 1 September 1943. Leopold Archives, ser. 9/25/10–1, box 3.
5. FNH to FH from Randolph Field, Texas, undated.
6. FNH, to fellow students, 14 December 1943, Leopold Archives, ser. 9/25/10–1, box 3.
7. FNH. Undated carbon copy of lecture notes preserved by Hammy.
8. Richard Eberhart, "The Fury of Aerial Bombardment," in *The Norton Anthology of Modern Poetry,* ed. Richard Ellmann and Robert O'Clair (New York: Norton 1973) 665, 665n. Eberhart wrote the first stanzas in a depressed state after reading the death lists that came regularly back to his base. It was only some years later that he was able to write the compelling last verse here quoted. Hammy, to my knowledge, never spoke in later years of his state of mind in 1944.
9. FNH to Commanding Officer, 29 October 1945.
10. FNH to Commanding Officer, 23 November 1945.
11. Elva Hamerstrom Paulson found this letter after Fran's death, thus casting some doubt on the family's report of a disappearance without trace. The puzzlement or irritation may have come from their leaving two active youngsters with aged parents at the time of Mr. Flint's illness. In fact, Mr. Flint died the next year, in September 1946.
12. Interview with Dagmar Lorenz Quatrini in her apartment in Florence, November 1990.

### Chapter 8. The Action

1. Some years after these events, Walter Bock (then at Columbia) and Keir Sterling (then at the University of Wisconsin, Madison) gathered information for a history of the AOU. After hearing from Mrs. Mayr, they contacted Fran. Her eight-page typescript was to have been part of an as yet unpublished book. I am grateful to Mr. Sterling for providing her 1983 typescript.
2. Fräulein Taggesell, to Fran, undated. She admonished, "I hope you are a good housewife and those duties must come before your hobby. . . . What are all those insects compared to a comfortable home, regular well prepared meals."

3. FH to Gustav Kramer, 23 January, 19 February 1947. Kramer filed the correspondence, and after his death his secretary found it and gave it to his daughter, who copied it for me.

4. FH to Erwin Stresemann, 9 February 1947.

5. Kramer to FH, undated.

6. FH to Kramer, 17 September 1948.

7. Helen Flint (Posey) mailed a letter from Pinckney to CARE on 11 June 1947. She asked that packages she had ordered for Fräuta be sent to others, thus establishing that the family governess died sometime before that date.

8. FH to Kramer, 16 April 1947. Detailed and excited observations about prairie chickens complete this letter. "Also after all these years we do not really know why they boom. We have produced booming in hens by putting testosterone propionate under the skin of the hind neck, but that still does not explain why they boom. Is it for intimidation? They boom at crow (*Corvus*), robins (*Turdus*), etc. They boom all alone when there are no other birds near. You have a rather similar display in some of your grouse don't you."

9. Ernst Mayr to author, 27 May 1992.

10. Gretel Mayr to Walter Bock, March 1982.

11. Gretel Mayr to FH, 18 June, probably 1947.

12. Ernst Mayr to author, 27 May 1992.

13. Kramer to Fran, from undated draft included in his papers.

14. I was puzzled by Fran's repeated statements to this effect, but the Adalbert Ebner file (see chapter 9) has lessened my perplexity. It is quite likely in those years that helping an "accused Nazi" would have brought threats of consequences.

15. FNH to Kramer, 5 August 1947.

16. Kramer to the Hamerstroms, 9 September 1947.

17. FH to Kramer, 9 November 1947.

18. Elizabeth Kramer to author, "Easter," 1992.

19. FNH to Kramer, 15 July 1947.

20. Kramer to FNH, 13 August 1947.

21. FH to Ernst Mayr, 25 October 1947. Unable to send to young ornithologists up till then, she described efforts to get more bird clubs and AOU committees involved.

22. Kramer to the Hamerstroms, 14 August 1947.

23. FNH to Kramer, 23 July 1948.

24. Anne-Marie Hershkovits to Kramer, 18 September 1947 from the Chicago Natural History Museum.

25. FH to Kramer, 18 December 1947.

26. Margaret Morse Nice, *Research Is a Passion with Me* (Toronto, Ontario: Consolidated Amethyst Communications, 1979), 249. She credited Fran with significant accomplishments in the Action, and for the auctions of

paintings that netted over $1,500 for the artists.

27. FNH to Kramers, 6 January 1948.

28. FH to Kramers, 26 February 1948.

29. FH to Kramers, 13 March 1948.

30. FH to "Gustav and Neni" Kramer, 20 April 1948.

31. FH to Kramer, 19 May 1948.

32. More formal records of this effort can be found in *The Auk* 66 (January 1949): 63. The small portion of the correspondence to which I have had access demonstrates the patience and perseverance the effort required. Seventeen Hamerstrom letters to the Kramers, for 1947 alone, survive—a fraction of the many that must have been written by others. The good will, humor, consideration, and ingenuity displayed on both sides are heartening, in this time of increased pain in the world. I was struck by the many tributes to others in the Hamerstrom letters, and the modesty with which all parties, repeatedly, belittled their own efforts.

### Chapter 9. The Return

1. Alexander Smith, *The Mushroom Hunter's Field Guide* (Ann Arbor: University of Michigan Press: 1958).

2. This material is in the glass case of the library of the Department of Wildlife Ecology at the University of Wisconsin–Madison.

3. Alan Beske, personal communication.

4. The agency changed its name to Department of Natural Resources (DNR) in the late 1960s. I use the more familiar, later designation in most citations except for the very early years.

5. FNH to Kramer, 27 August 1948.

6. FNH to Dr. J. S. Rogers, 22 December 1948. Rogers was director of the University of Michigan's Museum of Zoology.

7. FNH to Dean Hayward Keniston, 23 December 1948.

8. Art Staebler to FNH, 10 December 1948, included handwritten notes from colleagues.

9. FNH to Dr. Dansereau, 23 March 1954.

10. FH to Kramer, 2 November 1948.

11. FH to Kramer, 11 November 1948.

12. FNH to Kramer, 15 January 1949.

13. FH and FNH to the Kramers, 5 February 1949.

14. Aldo Leopold to Elizabeth Sammer (Ebner's sister), 10 November 1947.

15. FNH to Tom Gill, 16 June 1949.

16. Adalbert Ebner to FH, 12 October 1950. He felt compatible with them and recalled late-into-the-night conversations, indicating intense involvement during those graduate school days.

17. Sadaya Shimamoto to FNH, 1 February 1949.

18. AL to Elizabeth Sammer, 10 November 1947. Leopold later wrote (to Tom Gill, 3 February 1948) that he would feel betrayed were the charges true. My research found that a "Dr. E." is mentioned in the book, *Nazis in the Woodpile: Hitler's Plot For Essential Raw Material* (Bobbs-Merrill Company, New York: 1942), 158, as a German who spent considerable time in the United States. The author, Egon Glesinger, was the son of American businessman who owned forest property in Europe. He received a Ph.D from the Graduate Institute for International Studies in Geneva and became the secretary general of a *Comite International du Bois,* headquartered in Vienna. His book's goal was to acquaint Americans with a proposed German Wood Trust that tried to control European and world forest industries to provide fuel, food, textiles, structural materials, and chemical by-products for the coming war effort. Goering, its sponsor, hoped to build support for the Nazi plan among American foresters. Glesinger traces the failed diplomatic and commercial history of attempts to form that trust from 1932 until 1938 when he was advised to move the CIB from Vienna. He suspected his Austrian secretary, who went with him to his new location in Belgium, of being a spy and he mentions a forester, a Dr. E. who came to America for a long stay and then spent time in Japan. His conjectures are brief and vague; his primary focus is on the dangerous, but foiled policy of a worldwide lumber cartel.

The topic of German infiltration into high circles of the North American forestry establishment is larger than my subject, or resources. Leopold (see CM, *Aldo Leopold,* 351–360) was aware of Nazi cruelty well before the war.

19. FNH to Max Rheinstein, 7 March 1948.

20. Karl Reinhardt to AL, 23 April 1948.

21. FNH to Tom Gill, 12 December 1949.

22. See note 16 above.

23. Reid F. Murray, M. C. to High Commissioner McCloy, 6 December 1949. Hammy offered to pay the cost of the cable; the congressman took him up on the offer.

24. Ebner to FH and FNH, 12 October 1950.

25. The adder appears in *Strictly for the Chickens,* as one of the wild animals Fran has handled.

26. Ebner to FH and FNH, 12 October 1950.

27. Aldo Leopold, *Game Survey of the North Central States* (Madison: Sporting Arms and Ammunition Manufacturers' Institute, 1931), 161–162.

### Chapter 10. The Setting, the Task

1. Today, widespread cutting of Depression-era shelterbelts has enlarged fields to suit large tractors. Spring wind still scours emerging plantings.

Farmers wince when the grainy earth starts to skitter and cut the tender stems and emerging green shoots of cucumbers, peas, corn, and the pale, elongated sprouts of potatoes. Whipped seedlings, buried plants, and half-filled ditches still ensue.

2. Fran frowned as she read my copy: "You think it dismal. Hammy and I loved it!"

3. *Necedah Republican,* 5 April 1934.

4. CM, *Aldo Leopold,* 344.

5. According to the Geological Land Office in 1852, as a jotting on the back of an old township map in Hammy's file indicates.

6. "Report of the Waushara County Land Planning Committee," Fred Weymouth, Chairman, (33 pages, undated, mimeographed).

7. Dick Corey, personal communication.

8. AL, *Game Survey of the North Central States,* 161–188. The species originally inhabited northern Kentucky, Ohio, southwest Indiana, central and northeast Illinois, Iowa, and Western Missouri—prairie country.

9. Ibid., 171.

10. He predicted extinction by 1967.

11. Kabat, trained by Leopold, encouraged his people to get the Ph.D. degree. Many did: alumni include Larry Jahn, onetime head of the Wildlife Management Institute, and Ruth Hine, Don Thompson, Dick Hunt, and Dan Trainer (later dean of the College of Natural Resources at University of Wisconsin–Stevens Point).

12. AL to Bill Aberg, 13 March 1942. Leopold Archives, ser. 9/25/10–1, box 001.

13. FH to Kramers, 12 March 1950.

14. Oswald Mattson, personal communication.

15. Elva finished first grade in the one-room schoolhouse in Michigan, but was judged too young for second grade in Milton because she could not identify words like *trolley* or *fire hydrant.* Back in Plainfield she went directly to the second grade.

16. To reach visiting hunters he also put a story in the *Plainfield Sun,* 20 September 1951.

17. Woerpel lived in Stevens Point for years and was influential in preventing the damming of the Little Eau Pleine River, which allowed the Mead Wildlife Area to be established. *Stevens Point Daily Journal,* 9 February 1949.

18. FH, *My Double Life,* 190.

19. FNH to D. R. Thompson, 16 February 1955.

20. FNH to Kabat, 29 January 1953.

21. The mapping folder contains meticulously hand-drawn schematic maps from 1951–1965.

22. FNH to Ralph Hovind, 3 May 1951.

23. FNH to Elmer J. Haas, 19 August 1950.

24. Dick Hunt, personal communication.

25. FNH to Royal Henson, 7 September 1950.

26. Fran's list appears in the dedication to *My Double Life.*

27. Their survey established the richness of species in the Little Eau Pleine River area. In 1959 it was donated to the state. See George Rogers, *Stevens Point Journal,* 29 August 1984.

28. The author, Gordon MacQuarrie, became a lifelong champion.

29. *Milwaukee Journal,* 15 January 1950, 6. Fran was not unique: Eleanor Roosevelt, another active aristocrat, did not know who Frank Sinatra was.

### Chapter 11. Booming Chickens and a Land Boom

1. Gordon MacQuarrie, *Milwaukee Journal,* 15 January 1950.

2. Walter Scott founded the *Passenger Pigeon,* the magazine of the Wisconsin Society for Ornithology, along with many other services to the conservation cause.

3. Walter Scott to "Fred and Fran," 21 March 1953.

4. H. D. Ruhl to the DNR, 20 October 1953.

5. FNH to Andy Ammann, 16 April 1951.

6. Vilas Waterman to FNH, 24 February 1950.

7. Ruth Hine, who moved into editing, was hired in June as a conservation aide; Fran was hired for fieldwork later that fall.

8. Before I met the Hamerstroms, I heard: "There's a couple of scientists around here who live in luxury . . ." the indignant story began, "who even have a maid! and a *butler*!" This was the height of degeneracy. My surprise, when I walked into their house, can be imagined.

9. *Stevens Point Daily Journal,* 9 February 1949.

10. FNH to Kabat, 12 November 1953.

11. Signatures came from towns far from the marsh and the Milwaukee area. There were 85 from Wisconsin Rapids, 150 from Stevens Point and 59 from Plover, where druggist Gwidt held forth. In contrast, 8,000 petition signatures went to the Conservation Commission against Wallace and Hazel Grange's deer fence on their Wood County game farm in the early 1940s. See Grange, *Live Arrival Guaranteed,* 269–300.

12. Harry Walker to Ernie Swift, 28 November 1953.

13. Mike Primising, WCD staff in Waushara County, personal communication.

14. Grange, *Wisconsin Grouse Problems,* 268.

15. Eugene Parfitt to FNH, 2 February 1953.

16. FNH to Dick Deerwester, 29 December 1953. His letter credited Russ Neugebauer for the pricing advice, but quipped, "(Always give credit where

credit is due, and even where it isn't)." He owed Neugebauer no favors. See chapter 12.

17. She became Dory Vallier after Kummer's death shortly after this sale. Her second husband also supported buying land. See Tom Davis, "Dorothy and Jaques Vallier," in *Wisconsin Tales and Trails,* 30 no. 6 (Nov./Dec. 1989): 46. Dory told me that the Hamerstroms usually stayed in their comfortable guest room when they were in Milwaukee, but that "they always used only one of the single beds."

18. Frances Hamerstrom, "Management of Raptors," *Management of Raptors. Proceedings of the Conference on Raptor Conservation Techniques* 22–24, no. 2 (1973): 5–8.

19. FNH to S. C. Whitlock, Michigan Department of Conservation, 25 January 1954.

20. FNH to J. R. Smith, 16 November 1954.

21. Hammy was discreet, but a later letter to "Dave," (1 November 1980) was explicit. "When we first made it plain that chicken management depended on land control, the Department refused to purchase. It was the private sector that pulled our chestnuts out of the fire."

22. FNH to Deerwester, 5 October 1954.

23. Credit for originating the foundation idea is uncertain, but the Hamerstroms were without doubt responsible for its growth.

24. Hickey to members of the WSO, draft of proposed letter, sent to FNH, 16 November 1954.

25. Newsletter of the Dane County Conservation League. Undated, but probably 1976.

26. Paul Olson, personal communication.

27. Ray Anderson, personal communication.

28. "Dorothy and Jacques Vallier," *Wisconsin Tales and Trails,* 46.

29. Paul Hayes, *Milwaukee Journal,* 15 July 1990.

30. This comment became a byword in Hamerstrom circles.

31. Doug West to Morris Johnson, 11 April 1962.

32. Albert J. Carr, personal communication, 10 October 2000.

33. FNH in "BOOM!" (Society of Tympanuchus Cupido Pinnatus), 26 January 1967.

34. FH, *My Double Life,* 202–203.

35. Harry Croy, personal communication.

36. From John Emlen's draft of his obituary of Hammy. He sent it to Fran before submitting. On it, she wrote, "Johnny, it is just perfect!"

### Chapter 12. The Prairie Chicken War

1. FNH to Kabat, 22 June 1952. This memo is adapted for my account of the meeting.

2. Russ Neugebauer to Kabat and J. R. Smith, 2 November 1953.

3. The Portage county agent, M. P. Pinkerton, a jovial and politically astute man, consistently stood with the farmers during the controversy. Later he was on the unsuccessful Joint Management Committee, which successfully stonewalled any cooperative effort.

4. R. Neugebauer to FNH and FH, 12 November 1953.

5. FH, *My Double Life,* 264–266.

6. FNH to Larry Jahn, 7 October 1954.

7. FNH to J. R. Smith, 26 November 1954. Jim Hale told me that the Damon family, "denied project researchers access. I watched the booming ground on their land by peering over the top of the drainage ditch bank that was their south property boundary."

8. FNH, as quoted in the *Stevens Point Journal,* 17 June 1961.

9. *Milwaukee Sentinel,* 11 December 1954.

10. Cy Kabat to FNH, 8 April 1955.

11. FNH to H. H. Hovland, 31 October 1955.

12. *Stevens Point Journal,* 6 December 1954. The crowd of 150 at the Conservation Fraternity meeting at Central State College was large for that time.

13. "News and Views," (Wisconsin Federation of Conservation Clubs), December 1955.

14. *Plainfield Sun,* 30 December 1954.

15. Highway 51 cut through his land and was often obscured by blowing peat in the spring. I drove it, almost daily, and in 1992, when I left the area, patches of white sand gleamed from the formerly solidly black fields.

16. FNH to Kabat, 11 February 1955.

17. These were H. O. Schneiders, Al Walters, Gerry Vogelsang, and Dick Deerwester.

18. R. F. Deerwester's report, "Booming Ground Fine Shots," to unnamed parties, 23 February 1955 went on to say that "the letter sounded . . . like a 'brain washing' by the Department and Commission in an attempt to justify their past complacency and poor judgment."

19. FNH to Kabat, 16 May 55, 3 pages. "Our own work, both here and for the University, came just at the cyclic high and during the beginning of the downswing. Grange's work was during the downswing. . . . No one has followed a complete cycle through. *No one has studied the rising phase of the cycle.* The need for a better understanding of . . . the upswing is too obvious to require elaboration."

20. Kabat to J. R. Smith, 12 May 1955. Colleagues testify that without Kabat's diligent—even stubborn—efforts on their behalf, their project might well have come to grief.

21. J. R. Smith to Kabat, 17 May 1955; Kabat to J. R. Smith, 3 June 1955. Smith had said that he saw no reason for the Hamerstroms to live in the area.

22. FNH to J. R. Smith, 26 June 1955.

23. David Duffy, *Milwaukee Sentinel,* 6 November 1955. Their collecting permits allowed them to shoot specimens legally. The game warden found a warm partridge in the back of an Illinois car parked in their yard and did not, perhaps, believe the car's owner, who maintained that he had shot it in Wood County, a few miles west of the Hamerstrom farm where the season was still open. Nor did the warden consider that the bird would not have cooled in the heated car.

The DNR had tried to mute some problem at Horicon, but Duffey's suggestion that that matter had something to do with the Hamerstroms was simply false. They hired a good lawyer: the settlement of $2,000 financed their trip to Finland the next summer.

24. The date of the report was 10 August 1955. Hammy's penciled note was filed with it.

25. Owen Gromme was the curator of birds and mammals at the Milwaukee Public Museum. He was alert to protection of birds and supportive of such moves as flooding Horicon marsh and establishing the International Crane Foundation. His reputation rests on his very popular bird illustrations.

26. *Madison Wisconsin State Journal,* 23 January 1955.

27. Fourteen states were so designated; hunting was then permitted in only two of them, indicating the state of prairie chicken habitat at that time.

28. FNH to R. A. Brown, 24 January 1955.

29. One particularly forceful one, from Vogelsang to Guido Rahr, displays a penciled notation, "REV" in Hammy's hand, indicating that he played a part in writing it. MacQuarrie, *Milwaukee Journal,* 16 August 1955.

30. Kabat to FNH, 26 October 1955.

31. Wallace Grange to FNH, 18 August 1955.

32. Robert Ellarson, to the DNR, 26 March 1956. The committee rejected outright the idea that the prairie chicken area could be publicized as a tourist attraction, with sportsmen acting as guides, and dragged their feet on the food-patch proposal. Landowners wanted "no strangers trespassing on their lands." Ellarson, a respected UW Extension professor, took minutes of the meeting and sent them to Hammy with a confidential attachment: "A feeling of active belligerence seemed to pervade the meeting. . . . The group of farmers present . . . had absolutely no interest whatsoever in preserving the prairie chicken. Their interests seemed to be entirely selfish."

33. Response sheets, in Hammy's 1981 file.

34. FNH, undated carbon.

### Chapter 13. Hamerstroms' Kingdom

1. Joe Hickey to John T. Ratti, 17 August 1984, nominating Hammy for the Leopold medal. Hickey was then professor emeritus of Wildlife Ecology.

2. FNH to Morris Johnson, Fish and Game Department, Bismarck, N.D., 31 May 1962.

3. She could not afford to give the impression that she stinted her three-quarter-time employment, nor did she want her activities to cause any criticism of the DNR.

4. FH, *My Double Life,* 68–72. For the uninitiated, an austringer is a keeper of goshawks, a jess is a short strap for the captive hawk's leg, a creance is a fine line to hold a hawk during training, and musket is the name for a male sparrow hawk.

5. Hammy's Report to the DNR of 1 January 1957 cited numerous earlier studies to craft a careful analysis of Missouri range condition and potential, clearly intending to support wider extension of the scatter-pattern preservation plan.

6. Bruce Waage, then the chairman of the Prairie Grouse Technical Council, loaned me three boxes of correspondence in 1991.

7. Don Christisen to author, 24 May 1991.

8. Ibid.

9. FNH to Christisen, 17 May 1962.

10. FNH to Jerry Kobriger, 14 June 1971.

11. James B. Hale, personal communication. Hale contributed significantly both to planning and in applying for this grant, accompanying Hammy to Washington to assess requirements.

12. Ray Anderson, personal communication.

13. An instructive day that I spent with Dan at Cedar Grove substantially enriched my knowledge of both hawking and Hamerstrom activities.

14. R. C. Hopkins to FNH, 29 March 1960. That year, 21 percent of all land in Oasis township was in the program, a considerable acreage to record.

15. R. G. Lynch, *Milwaukee Journal,* 29 October 1961.

16. FNH to Matson and Stan de Boer, 3 May 1962.

17. Stanley de Boer, mimeographed letter to "Dear Conservationist," 9 May 1962.

April 11, 2002.

18. FNH to Mattson and Ben Hubbard, 11 September 1966.

19. He insisted that flexibility would be required as land use changed. Were plowland to appear, their plan would have to be modified with larger self-supporting units.

20. Os Mattson, "Wisconsin Prairie Chicken Management, Progress and Recommendations" was presented orally.

21. FNH to Hale, 4 January 1965.

22. Hale to Mattson, 8 January 1965.

23. Rigging a drop net is exacting. After trying to put up a new drop net that would not hang in the required stretched position, Hammy wrote an unidentified "Bob," 10 May 1965: "When we put the net up the first time it

bagged to the ground. . . . Fran took it back and took a crew of men out into the city park where they had room enough to spread it flat and hang it properly. The trick is a mathematical one: take the actual measurement of the mesh . . . and measure or calculate the diagonal of one mesh when it is laid out square. Divide the width of the net by the diagonal of the mesh size to find out how many meshes per side. The net should be hung on the bias, as a fish net is."

24. FNH to Christisen, 6 June 1963.

25. FNH to Hickey, 16 August 1963.

26. *Stevens Point Daily Journal,* 29 August 1958.

27. FNH to an unidentified Harry [Croy?], 8 January 1965.

28. FNH to Hale, 10 August 1970.

29. Michael Tlusty of Syracuse University noted that Hammy felt constrained by the rigidity of scientific protocols that prevented investigation of unusual and interesting "side events." *Journal of Raptor Research* 26, no. 3 (1992): 195.

30. FNH to J. R. Smith, 8 July 1964.

31. FNH to Christisen, 26 October 1963.

32. Mattson to FNH, 8 May 1964.

33. *Rural Electric Cooperative News,* 8 March 1970.

34. FNH to Hale, 29 February 1964.

35. A letter from Mattson to Dan Thompson (28 August 1964) may indicate a substantial reason for the move: "I told Bill [Sullivan?]—we just couldn't promise to tackle more land with present manpower and equipment until we get caught up on our present lands. I know he will make a pitch for more help and equipment—I also know this won't help cement Dept. relations but I can't help it. . . . I want more land badly and I want to see it developed as soon as Bill does but it just can't be done with 2 men, second rate equip, inferior herbicides, unfriendly ranger etc."

36. FNH to Christisen, 12 February 1965.

37. FNH to Christisen, 4 June 1965.

38. FNH, Annual Report, 30 June 1965.

39. Ibid.

40. FNH to Westemeier, 7 July 1970.

41. FNH to Hale, 25 May 1968.

42. FNH to Thomas L. Kimball, 25 March 1971.

43. *Stevens Point Journal,* 12 May 1971. Clarence Bernhagen was the farmer.

### Chapter 14. A Naturalist Family

1. One of our neighbors returned from a 1960 UW Extension presentation on child rearing, dissatisfied. "They wouldn't even tell me," she

grumbled, "when to spank my children!" And a Plainfield High gym teacher once whispered to me, "You know the daughter of those bird people? That girl is *clean!*" I was astonished at her surprise.

2. Their paper, "Comparison of European Black Grouse and Prairie Chicken—Their Mating Display," became part of a later publication: FNH and FH, "Comparability of Some Social Displays of Grouse," *Proceedings of the International Ornithological Congress* 12 (1960): 274–293.

3. Peter entertained us in his home in Gland, Switzerland, and Lorenz came to Eli's apartment in Weinheim, Germany, to see us. See chapter 7, note 12.

4. Elva agreed. "Posey's efforts to bring order to our house drove Fran and Hammy absolutely bats. They would never have arranged such long visits unless they had really needed her during hunting season. But Hammy was always considerate." Fran said, "Hammy treated Posey like a princess."

5. Fran's report was that the wire said, "Please let her stay. It might be the best for her to be in America rather than with weeping Aunts." It is hard to believe that in those circumstances Hammy would have allowed so abrupt a wording.

6. A. Ebner to "Fran and Hammy Hamerstrom," 9 March 1959.

### Chapter 15. Of Hawks, Humans, and Freedom

1. Their experience with the Prairie Chicken Foundation became a model as Butch Olendorf, a longtime member of the Raptor Research Foundation acknowledged. "I could always call him. With his advice and a lot of work we managed to stay in business." (Fran was a director in 1975 and 1976.) Personal communication. Fran was president of the Wisconsin Society of Ornithologists in 1960, Hammy in 1966 and 1975. He served on several committees; they cochaired the Research Committee from 1962 to 1979.

2. Ruth Hine, personal communication.

3. FNH to Robert Evans, 20 September 1962.

4. He was on the editorial board of the *Journal of Raptor Research* for almost fifteen years, as Principal Referee from 1975 on, and edited *Management of Raptors,* the *Proceedings* of 1974. He was associate editor for the Wilson Ornithological Society from 1942 to 1948; he served on the editorial board of the Wildlife Society from 1957 to 1972. No count can ever be made of the student papers he improved and of the learnings that occurred under his tutelage.

5. Fran told me that he won the Edward's prize for a paper on harriers in the prestigious *Wilson Bulletin* thus: "He pulled together my raptor data, and hired a gabboon to do the statistics." Their bibliographies, however, record him as the senior author, with Fran and Charles Burke, for "Effect of Voles on Mating Systems in a Central Wisconsin Population of Harriers,"

*Wilson Bulletin* 97 no. 3 (1985): 332–346. Fran, angered by the opinion that he should have produced more papers, constantly defended his production.

6. Ray Anderson, in a tribute delivered at the 1993 ceremony installing the Hamerstroms in the Conservation Hall of Fame, Stevens Point, Wisconsin.

7. Dan Berger, personal communication.

8. Frances Hamerstrom, *An Eagle to the Sky* (Ames: Iowa State University Press, 1970). Its early chapters had previously appeared in *Natural History*.

9. FH, *Harrier, Hawk of the Marshes,* 19. She speaks of the "exhilarating sight of captured hawks."

10. FH, *Birding with a Purpose,* 36. She also credited Dan Berger and Helmut Mueller for modernizing the bal-chatri so that it could be used anywhere.

11. FH, *An Eagle to the Sky,* 49–50.

12. FH, *Birding with a Purpose,* 49.

13. Ibid. "On the art of Harris' Hawk Trapping" and "Winter Visitors from the Far North" illustrate a variety of applications.

14. The Milwaukee Public Museum had banded hawks at Cedar Grove on Lake Michigan near Sheboygan since 1935. Bird-watcher Dixie Larkin and Wisconsin wildlife artist Owen Gromme persuaded the Conservation Department to buy thirty-two acres surrounding the small Civilian Conservation Corps project in 1948. They organized volunteers to study raptor migration on this major flyway. Work stopped during the war; at its end Helmut Mueller and Dan Berger reactivated the project. Volunteers record sightings of migrating raptors from dawn to dusk each spring and fall, trapping, banding, measuring, and releasing them.

15. The archives are located at the World Center for Birds of Prey on the outskirts of Boise, Idaho.

16. That map, updated, now graces the cover of the falconers' association newsletter. The entire continental United States is in black; Hawaii, with only one endemic raptor, the Hawaiian hawk, on the endangered list, forbids any import of exotics, so the sport cannot be practiced in the islands.

17. It was likely in recognition of this effort that she received the President's Award from the Raptor Research Foundation in 1975, the year she served as central director. Alan Beske, "Dr. Frances Hamerstrom: 1907–1998," *North American Falconers' Association* 37 (1998): 120. Fran also received the Heritage Award of the Archives of American Falconry and NAFA the week before her death in 1998; Carnie was able to speak to her about it.

18. FH, *Harrier, Hawk of the Marshes,* 141.

19. Fran's account in *Harrier, Hawk of the Marshes,* 20–21 differs considerably from Dan Berger's. He is sure that she learned to use dho-gazas at Cedar Grove and borrowed used ones from him for the harrier project, at least until 1960. It is probable that Fran did look up trapping techniques in Christian Ludwig Brehm's 1855 book as she said, but she may well have

been careless in recording the time that she did so. Fran valued chronology much less than do many people.

The reader interested in the technique may find information in Peter H. Bloom et al., "The Dho-Gaza with Great Horned Owl Lure: An Analysis of Its Effectiveness in Capturing Raptors," *Journal of Raptor Research* 26, no. 3 (September 1992): 167–179. Its final paragraph salutes Fran and Frederick Hamerstrom "for bringing this technique back from the many centuries it lay essentially dormant in the Old World."

20. FH, *Harrier, Hawk of the Marshes,* 57.

21. Ibid., 57–59. That number would seem to a lay person to involve enormous labor, but the amount was small compared to the vole index run by Cornell for three years, which involved surveying and counting of grass stems cut by the voles.

22. Ibid., 49–50.

23. Joseph J. Hickey, ed., *Peregrine Falcon Populations: Their Biology and Decline* (Madison: University of Wisconsin Press, 1969).

24. Frances Hamerstrom, "Management of Raptors," *Management of Raptors. Proceedings of the Conference on Raptor Conservation Techniques* 22–24, no. 2 (1973): 8.

25. Deann became a wildlife artist. The strength of her connection to the Hamerstroms became clear when she chose to live with, and sensitively care for Fran during the latter's demanding final years.

26. See "Short Communications" and "Letters," *The Journal of Raptor Research* 26, no. 3 (1992): 89–210, This celebration issue, with a sensitive tribute to the pair by Ernst Mayr, was edited by Josef K. Schmutz in association with several ex-gabboons and friends of the Hamerstroms. It is a record of their impact.

27. Fran thought every child should have a wild pet—a snake, a frog, even an insect—as a valuable introduction to nature study. She built upon her own experience but also adopted Leopold's suggestion that rearing a wild animal in captivity was a valuable precondition for wildlife management training. She drew herself up one day, when I questioned the utility of her recommendation in today's world, and said, "Leopold said that animals, plants, and soils are the *alphabet* of wildlife management!"

28. Deann de la Ronde to author, 15 August 2000.

29. A fine brief example of Hammy's rigor is "An Ecological Appraisal of the Peregrine Decline," in Hickey, *Peregrine Falcon Populations,* 509–513. His tone is conversational. "What are we looking for?"; his analysis subtle and complete. Common sense prevails. "We got on top of a beautiful mountain, and he took a deep breath and said, 'Wonderful! Wilderness! Nobody here!' There was no one else there at the moment, but one had only to look at any rock to see the scratches of hobnails that had been there before." His survey of the anomalies in the population situation in 1965 demonstrates his

masterly use of queries and supposition, and the research leads derived from them. Were more scholarly writing like Frederick Hamerstrom's, today's science would not be so far removed from the ordinary human.

30. Fran lost the sight of one eye in the late 1960s.

31. William C. Scharf, "Letters," *Journal of Raptor Research* 26, no. 3 (1992): 199.

32. Joe Schmutz, personal communication.

33. Dale Gawlik, "Conservation and the Land Ethic," *Journal of Raptor Research* 26, no. 3 (1992): 181.

34. Ruth Hine, personal communication.

35. FNH to Hale and W. A. Creed, 7 October 1969. This memo reveals that he was now reporting progress and delays on five areas, from taxonomy and habitat to hybrids. He asked for help: "This summer's cover mapping has added important data. The Buena Vista fieldwork is done, and I expect to finish Leola this afternoon. Buena Vista has been dot counted . . . the new map needs to be drafted (I trust by Engineering). Both dot counting and drafting are needed for Leola; we have data taken at four times from 1941 through 1969. We could use help here."

36. Don Thompson, personal communication.

37. FN To "Ingemar" [Hjorth], 21 March 1969. Hjorth was a young European ornithologist they had befriended. See chapter 16, note 11.

38. FNH to Jerry [Kobriger?], 14 June 71.

39. The letter she spoke of was dated 19 September 1971.

40. James A. Crowe, Outdoors Editor, *Detroit Sunday News*, 19 December 1971.

41. FNH to Kabat, 29 August 1973.

42. *The Prairie Chicken in Wisconsin: Highlights of a 22-Year Study of Counts, Behavior, Movements, Turnover, and Habitat,* Technical Bulletin 64 (Madison: Wisconsin Department of Natural Resources, 1973.); FNH to Kabat, 29 August 1973.

43. *Stevens Point Daily Journal,* 13 August 1973.

### Chapter 16. Free at Last

1. Don Christisen to author, 21 September 1991.

2. FNH to Pat Carr, 2 June 1977. In 1999, I discovered that my new neighbor in New Mexico, Pat Carr Drypolcher, had written this paper in high school. She and her parents knew the Hamerstroms well.

3. FNH to R. Westemeier, 9 October 1978.

4. Westemeier to FNH, 4 May 1977.

5. FNH to Glen Sanderson, 12 March 1979.

6. Konrad Lorenz, *King Solomon's Ring* (New York: Thomas Y. Crowell, 1952), xviii.

7. 1970 *An Eagle to the Sky,* 1972 *Birds of Prey of Wisconsin,* 1973 *Walk When the Moon Is Full,* 1977 *Adventure of the Stone Man,* 1980 *Strictly for the Chickens,* 1981 *Is She Coming Too?* 1984 *Birding with a Purpose,* 1985 *Wild Food Cookbook,* 1986 *Harrier, Hawk of the Marshes,* 1994 *My Double Life.*

8. Dennis Ribbens, "Bookmarks/Wisconsin," *Wisconsin Academy Review* 27, no. 3 (June 1981): 35–36.

9. John Miner, *Appleton Post Crescent,* 6 June 1971. By Hammy's reckoning, 2021 would be the time when the environment would be damaged beyond repair.

10. *St. Paul Pioneer Press,* undated clipping, probably from the early 1980s.

11. A single example of such help suffices: A young European ornithologist wanted to film North American grouse and sent them an elaborate itinerary. He intended to come on public transportation from Lewiston, Montana, direct to Plainfield, with separate shipping for his considerable gear. After much correspondence, delay, changes of schedule, and difficulties with luggage, he arrived with his wife and young daughter—in booming time. He brought his master's thesis with him and asked that they review its contents and his English. When his letter of thanks arrived, Fran answered with a poem; they remained friends for years.

12. With an unusual show of dislike, he told us, "He behaved as though he was on God's baseball team the way he slapped me on the shoulder." When he told church authorities of the situation, they put a stop to such sponsorships.

13. This account is based on personal communications from the Hamerstroms and Mr. and Mrs. Oar. For further coverage of the extent and results of this action, see Paul McKay, "Operation Falcon: A Special Report," *Whig Standard Magazine* (10 October 1987).

14. FNH, *Stevens Point Journal,* 20 September 1982.

15. From College Station, Texas, nearly to the border. Novy Silva's account differs. Fran called him for help when a raccoon killed her bait pigeons. He obliged; the drive took all of his Sunday.

16. Art Hawkins, personal communication.

17. FNH as quoted by Joyce Hubbard, "Reach," *Stevens Point Journal,* 1 February 1989.

18. Porfirio, at the age of twenty-four or twenty-five, was still alive in December 2000.

19. *Wisconsin Rapids Daily Tribune,* 7 August 1990.

20. Edith Nash, to her daughter, undated.

21. Thad Smith to Fred and Fran, 29 June 1981. Thad wanted to make sure that everyone knew of Hammy's principles. Hammy had admired Thad's playing tennis with one of the few black students at Dartmouth when not

many would have done so. "He stood for what his conscience told him was right and lived by it, never wavering or yielding."

22. FH to Putnam Flint, 16 July 1981.

### Chapter 17. The Making of a Legend

1. Mrs. Jim Teer, personal communication.
2. FH, *Is She Coming Too?* 116.
3. FH, *My Double Life,* 155.
4. Ibid., 137–140.
5. I heard this story long before I met the Hamerstroms from Dr. Longenecker.
6. They appreciated our loan of migrant worker Josie one summer. Josie much preferred cooking enchiladas for them to picking cucumbers on our farm. However, she inquired, after reporting that she had been firmly directed to wash the sterling separately, "I theenk, maybe, that lady a little bit crazy?"
7. This is an example of Fran's need to dramatize. Elva reports that Mrs. Chamberlain did laundry until she "keeled over with a heart attack."
8. Elizabeth Kramer (see chapter 14), herself free spirited, gleefully recalled this incident.
9. Late in life, they donated some of these skins to the Smithsonian Institution, delivering them on one of their trips east.
10. Ray Anderson, personal communication.
11. Dan Berger, personal communication
12. Ralph Hopkins and Mike Primising, interview with author. Mr. Primising was no longer sure of the proper name of the insect Hammy had identified.
13. FH, *My Double Life,* 233.
14. FH, *Harrier, Hawk of the Marshes,* 91.
15. Alan Hamerstrom to FH, 24 May 1991.
16. Roger Fouts with Stephen Tukel Mills, *Next of Kin: What Chimpanzees Have Taught Me about Who We Are* (New York: William Morrow, 1997).
17. Evelyn Fox Keller, *Reflections on Gender and Science* (New Haven: Yale University Press, 1985), 75. New definitions, 117.
18. AL, *A Sand County Almanac,* 240.
19. She once repeated a comment she attributed to Agassiz, that, even in his day, his students no longer knew how to keep an animal alive.
20. FH, *An Eagle to the Sky,* 7, 9, 11. She used the hot water bottle that "we had used to abate the miserable earaches of our son."
21. AL, *A Sand County Almanac,* 116.
22. Grange, *Live Arrival Guaranteed,* 349.

23. Vera John-Steiner, *Creative Collaboration* (New York: Oxford University Press, 2000), 14.

24. Ibid., 11.

25. Helena M. Pycior, Nancy G. Slac, and Pnina G. Abir-Am, eds., *Creative Couples in the Sciences* (New Brunswick, N.J.: Rutgers University Press, 1995). The references below are, in order: 181, 148, 151–153, 165–168.

26. AL, *A Sand County Almanac,* 116–119.

27. John-Steiner, "Flights and Resting Places: Discovering Complimentarity in Scientific Collaboration," lecture, Santa Fe Institute, 17 October 2000.

### Chapter 18. Death of a Biologist

1. Fran tried to keep all her books in stock. Were one remaindered, she bought the lot to sell at retail at every opportunity. Proceeds supported their research.

2. After Fran's death the Raptor Research Foundation took over the existing fund and established a memorial, the "Frederick and Frances Hamerstrom Award." Contributions are still received through the Society of Tympanuchus Cupido Pinnatus Ltd., Stone Ridge II—Suite 280, NW14W23777 Stone Ridge Drive, Waukesha, WI 53188–1188 or the Raptor Research Foundation, Inc., 14377 117th St. South, Hastings, MN 55033.

3. William Janz, *Milwaukee Journal Sentinel,* 29 April 1989.

4. Alan Hamerstrom, personal communication.

5. See Robert A. McCabe, "Frederick N. Hamerstrom 1909–1990," *Passenger Pigeon* 52, no. 2 (1990): 211–212. Others saw Hammy more clearly. Biologist Charles Long wrote Fran (6 August 1990) to say that Hammy didn't publish much was " pure big university elitism. . . .The measure of a scientist is overall productivity. Publication gives one subjective estimate of that. The measure of a naturalist is the love for nature. On that Fred gave full measure. Scientists are fine, I include myself among them, but genuine naturalists are gods."

6. Ray Anderson, "A Tribute to the Hamerstroms," *Abstracts, 18th Prairie Grouse Technical Council Conference* (N.p.: Michigan Department of Natural Resources, 1989), 25–30.

### Chapter 19. Fran

1. They would have. Fran spoke of her eye trouble in a letter to her mother from Falmouth in 1931; it was serious. Years later, leaving Wisconsin to go east for treatment, she said she wondered, as she said goodbye, if she would

ever see Hammy again. Her good eye weakened in her old age.

2. Sabine Strecker, "Letters," *Journal of Raptor Research* 26, no. 3 (1992): 198.

3. Gary Laib, personal communication.

4. Dale Gawkick, "In Memoriam: Frances Hamerstrom," *Journal of Raptor Research* 32, no. 4 (1998) ii–iv.

5. Peter Kramer kindly gave me permission to use this letter, written in 1996.

### Chapter 20. A Postscript

1. I spent a fine spring day (22 May 2001) with Jim at his ranch near Portales, N. M.

2. Later that day Jim confirmed for me that occasional remnants of the original prairie are found near Elida.

3. "Jim Weaver . . . set a standard and taught us why the effort was worthwhile." Dan O'Brien, *The Rites of Autumn: A Falconer's Journey across the American West* (New York: Atlantic Monthly Press, 1988), n.p. Weaver said, explicitly, "I always envisioned the Peregrine Fund working the same way the Society Tympanuchus Pinnatus Cupido did. It was the model: hard work, volunteerism, cooperation."

4. See Nathan F. Sayre, *The New Ranch Handbook: A Guide to Restoring Western Rangelands* (Santa Fe, N. M.: Quivera Coalition, 2001). Sustainable agriculture requires a management style that gives attention to the individual situations of each landscape, recognizes the complex ecosystems involved, and applies scientific research to each circumstance. Sharing results builds common vocabularies, thus making it easier for others to copy or adapt solutions found, and individuals of widely differing beliefs and mandates can start preventing and healing land degradation.

5. The herbicide of choice is tebuthiuron, previously used by the Bureau of Land Management in efforts to increase grass cover. See Martin Frentzel, "Hail to the Hens," *New Mexican Wildlife* 46, no. 1 (Spring 2001): 7–10.

6. The New Mexico Department of Game and Fish, The U.S. Fish and Wildlife Service, the Fish and Wildlife Foundation, El Llano Estacado Resource Conservation and Development Inc. are involved.

7. Conrad Richter, *The Sea of Grass* (New York: Alfred A. Knopf, 1936), 80.

8. Aldo Leopold, "The Farmer as Conservationist," from *The Leopold Outlook: Notes from "the Shack,"* newsletter of the Aldo Leopold Foundation, Inc., 2, no. 2 (Summer 2000): 2. The essay is available in *The Health of the Land,* eds. J. Baird Callicott and Eric T. Freyfogle (Washington, D.C.: Island Press, 1999.)

9. Weaver's account differs only slightly from Fran's in *An Eagle to the Sky,* 65.

10. Scott Weidensaul, "Stage Grouse Strut Their Stuff," *Smithsonian* 33, no. 3 (May 2001): 56–63.

11. W. Daniel Svedarsky, Ross H. Hier, and Nova J. Silvy, eds., *The Greater Prairie Chicken: A National Look,* Miscellaneous Publication 99–1999 (Saint Paul: Minnesota Agricultural Experiment Station, University of Minnesota, 1999).

12. "We should have started this much earlier," said Novy Silva to me recently. "We didn't get enough genetic variance in the stock from which we bred." He would like to see captive breeding commence on the lesser prairie chicken.

13. W. Daniel Svedarsky and Brian Winter, preface, *The Greater Prairie Chicken,* Svedarsky, Hier, and Silvy, eds., i.

14. I am well aware of the intensity of effort such as in John Toepfer's continuing work and the impressive papers published by Ron Westemeier in recent years, but any survey of current research is well beyond the intent of this book. See Svedarsky, Hier, and Silvy for further information.

15. Douglas H. Chadwick, "Down to a Handful," *National Geographic* (March 2002): 49–61. www.nationalgeographic.com/ngm/0203.

16. The Quivera Coalition (see note 4 above) reaches out in well-publicized, cooperative and practical ways to ranchers in many parts of the state.

17. Mary Riseley, "Ranchers and Environmentalists Collaborate: Learnings from a Corner of the Southwest," master's thesis, University of Massachusetts–Boston, 1998.

18. Dale E. Gawlik, "Conservation and the Land Ethic," *Journal of Raptor Research* 26, no. 3 (1992): 179–183.

19. Steve Lerner, *Eco-Pioneers: Practical Visionaries Solving Today's Environmental Problems* (Cambridge, Mass.: MIT Press, 1977).

# Index

341